George Douglas Campbell Argyll

Organic Evolution Cross Examined

Some Suggestions on the Great Secret of Biology

George Douglas Campbell Argyll

Organic Evolution Cross Examined
Some Suggestions on the Great Secret of Biology

ISBN/EAN: 9783744689564

Printed in Europe, USA, Canada, Australia, Japan

Cover: Foto ©berggeist007 / pixelio.de

More available books at **www.hansebooks.com**

ORGANIC EVOLUTION

CROSS-EXAMINED

OR SOME SUGGESTIONS ON THE GREAT SECRET OF BIOLOGY

BY THE

DUKE OF ARGYLL

K.G., ETC.

LONDON

JOHN MURRAY, ALBEMARLE STREET

BOSTON

LITTLE, BROWN AND CO.

1898

PREFACE

THE three Chapters in this work—little altered — were all originally contributions to the *Nineteenth Century*, which by the kind permission of the Editor, Mr. Knowles, I now republish in a separate and connected form.

Mr. Spencer, in the May 1897 number of the same Review, has ascribed to me, in these papers, several misconceptions as to his contentions and position. These, however, are all open to argument — except one. In this one Mr. Spencer thinks I have represented him as accepting a comparatively short period for the duration of the living world—whereas he merely argued that even assuming the shorter period, it might be quite long enough for the evolutions of Biology. I quite understood this, and have altered a few words to make the meaning clearer. In my reasoning, and in his former reasoning, everything turns not on the actual time, but on the supposed necessity for some enormous time. This is abandoned in Mr. Spencer's new argument, and the change is one having all the significance that I attach to it.

ARGYLL.

CHAPTER I

A GREAT CONFESSION

AMONG the many distinguished men who have contributed to the world's plebiscite in favour of the Darwinian hypothesis on the origin of species, there is no one so distinguished as Mr. Herbert Spencer. He alone has dealt with it systematically. He has pursued the idea of development with wonderful ingenuity through not a few of its thousand ramifications. He has carried it into philosophy and metaphysics. He has clothed it in numerous and subtle forms of speech, appealing to various faculties, and offering to each its appro-

B

priate objects of recognition.   He is the
author of that other phrase, "the survival
of the fittest," which has almost superseded Darwin's own original phrase of
" natural selection."   Nothing could be
happier than this invention for the
purpose of giving vogue to whatever it
might be supposed to mean.   There is
a roundness, neatness, and compactness
about it, which imparts to it all the
qualities of a projectile with immense
penetrating power.   It is a signal
illustration of itself.   It is the fittest of
all phrases to survive.   There is a sense
of self-evident truth about it which fills
us with satisfaction.   It may perhaps be
suspected sometimes of being a perfect
specimen of the knowledge that puffeth
up, because there is a suggestion about
it—not easily dismissed—that it is tautological.   The survival of the fittest may
be translated into the survival of that

which does actually survive. But the special power of it lies in this, that it sounds as if it expressed a true physical cause. It gets rid of that detestable reference to the analogies of mind which are inseparably associated with the phrase of natural selection. It is the great object of all true science—as some think—to eliminate these analogies, and if possible to abolish them. Survival of the fittest seems to tell us not only of that which is, but of that which must be. It breathes the very air of necessity and of demonstration. Among the influences which have tended to popularise the Darwinian hypothesis, and to give it the imposing air of a complete and satisfactory explanation of all phenomena, it may well be doubted whether anything has been more powerful than the wide acceptance of this simple formula of expression.

Such is the authority who some

years ago contributed to the *Nineteenth Century Review* two papers upon " The Factors in Organic Evolution." The plural title is significant. The survival of the fittest is a cause which after all does not stand alone. It is not so complete as it has been assumed to be. There are in organic evolution more elements than one. There is concerned in it not one cause but a plurality of causes. A "factor" is specially a doer. It is that which works and does. It is a word appropriated to the conception of an immediate, an efficient cause. And of these causes there are more than one. Neither natural selection nor survival of the fittest is of itself a sufficient explanation. They must be supplemented. There are other factors which must be admitted and confessed.

This is the first and most notable feature of Mr. Spencer's articles. But there is another closely connected with it, and that is the emphatic testimony he bears to the fact that the existing popular conception is unconscious of any defect or failing in the all-sufficiency of the Darwinian hypothesis. He speaks of the process brought into clear view by Mr. Darwin, and of those with whom he is about to argue, as men "who conclude that taken alone it accounts for organic evolution."[1] In order to make his own coming contention clearer, he devises new forms of expression for defining accurately the hypothesis of Darwin. He calls it "the natural selection of favourable variations." Again and again he emphasises the fact that these variations, according to the theory, were "spontaneous," and that their utility was only "fortunate," or, in other words, accidental. He speaks of them as

[1] P. 570.

" fortuitously arising " ;[1] and it is of this theory, so defined and rendered precise, that he admits it to be now commonly supposed to have been " the sole factor " in the origin of species.

It is surely worth considering for a moment the wonderful state of mind which this declaration discloses. When Mr. Herbert Spencer here speaks of the " popular " belief, he is not speaking of the mob. He is not referring to any mere superstition of the illiterate multitude. He is speaking of all ranks in the world of science. He is speaking of some overwhelming majority of those who are investigators of Nature in some one or other of her departments, and who are supposed generally to recognise, as a cardinal principle in science, that the reign of law is universal there—that nothing is fortuitous—that nothing is the

[1] P. 575.

result of accident.   Yet Mr. Herbert Spencer represents this great mass and variety of men as believing in the preservation of accidental variations as "the sole factor," and as the one adequate explanation in all the wonders of organic life.

Nor can there be any better proof of the strength of his impression upon this subject than to observe his own tone when he ventures to dissent.   He speaks, if not literally with bated breath, yet at least with a deferential reverence for the popular dogma, which is really a curious phenomenon in the history of thought. "We may fitly ask," he says, whether it "accounts for" organic evolution.   "On critically examining the evidence," he proceeds, "we shall find reason to think that it by no means explains all that has to be explained."   And then follows an allusion of curious significance.   "Omitting," says Mr. Spencer, "for the present any consideration of a factor which may be distinguished as primordial—"[1] Here we have the mind of this distinguished philosopher confessing to itself—as it were in a whisper and aside—that Darwin's ultimate conception of some primordial "breathing of the breath of life" is a conception which can only be omitted "for the present." Meanwhile he goes on with a special, and it must be confessed a most modest, suggestion of one other "factor" in addition to natural selection, which he thinks will remove many difficulties that remain unsolved when natural selection is taken by itself. But whilst great interest attaches to the fact that Mr. Herbert Spencer does not hold natural selection to be the sole factor in organic evolution, it is more than doubtful whether any value attaches to the new factor with

[1] P. 570.

which he desires to supplement it. It seems unaccountable indeed that Mr. Herbert Spencer should make so great a fuss about so small a matter as the effect of use and disuse of particular organs as a separate and a newly recognised factor in the development of varieties. That persistent disuse of any organ will occasion atrophy of the parts concerned is surely one of the best established of physiological facts. That organs thus enfeebled are transmitted by inheritance to offspring in a like condition of functional and structural decline is a correlated physiological doctrine not generally disputed. The converse case —of increased strength and development arising out of the habitual and healthy use of special organs, and of the transmission of these to offspring—is a case illustrated by many examples in the breeding of domestic animals. I do not

know to what else we can attribute the
long slender legs and bodies of grey-
hounds so manifestly adapted to speed
of foot, or the delicate powers of smell
in pointers and setters, or a dozen other
cases of modified structure effected by
artificial selection.

But the most remarkable feature in
the elaborate argument of Mr. Spencer
on this subject is its complete irrelevancy.
Natural selection is an elastic formula
under which this new "factor" may be
easily comprehended.   In truth the
whole argument raised in favour of
structural modification arising out of
functional use and disuse, is an argument
which implies that Mr. Spencer has not
himself entirely shaken off that interpretation of natural selection which he is
disputing.   He treats it as if it were the
definite expression of some true physical
and efficient cause, to which he only

claims to add some subsidiary help from another physical cause which is wholly separate. But if natural selection is a mere phrase, vague enough and wide enough to cover any number of the physical causes concerned in ordinary generation, then the whole of Mr. Spencer's laborious argument in favour of his "other factor" becomes an argument worse than superfluous. It is wholly fallacious in assuming that this "factor" and "natural selection" are at all exclusive of, or even separate from, each other. The factor thus assumed to be new is simply one of the subordinate cases of heredity. But heredity is the central idea of natural selection. Therefore natural selection includes and covers all the causes which can possibly operate through inheritance. There is thus no difficulty whatever in referring it to the same one factor whose solitary dominion Mr. Spencer has plucked up courage to dispute. He will never succeed in shaking its dictatorship by such a small rebellion. His little contention is like some bit of Bumbledom setting up for Home Rule — some parochial vestry claiming independence of a universal empire. It pretends to set up for itself in some fragment of an idea. But here is not even a fragment to boast of or to stand up for. His new factor in organic evolution has neither independence nor novelty. Mr. Spencer is able to quote himself as having mentioned it in his *Principles of Biology* published some twenty years ago ; and by a careful ransacking of Darwin he shows that the idea was familiar to and admitted by him at least in his last edition of the *Origin of Species*. Mr. Spencer insists that this fact is evidence of a "reaction" in Darwin's mind against the sole factorship of natural selection. Darwin was a man so much wiser than all his followers, and there are in his book so many indications of his sense of our great ignorance, that most probably he did grow in the consciousness of the necessary incompleteness and shortcomings of his own explanations. But there was nothing whatever to startle him in the idea of heredity propagating structural change, through functional use and disuse. This idea was not incongruous with his own more general conception. On the contrary, it was strictly congruous and harmoniously subordinate. He did not profess to account for all the varieties which emerge in organic forms. Provisionally, and merely for the convenience of leaving that subject open, he spoke of them as fortuitous. But to assume the really fortuitous or accidental character of variation to be an essential part of this theory is merely one of the many follies and fanaticisms of his followers.

Although, therefore, the particular case chosen by Mr. Herbert Spencer to illustrate the incompetency of natural selection, taken alone, to explain all the facts of organic evolution is a case of little or no value for the purpose, yet the attitude of mind into which he is thrown in the conduct of his argument leads him to results which are eminently instructive. The impulse "critically to examine" such a phrase as "natural selection" is in itself an impulse quite certain to be fruitful. The very origin of that impulse gives it of necessity right direction. Antagonism to a prevalent dogma so unreasoning as to set up such a mere phrase as the embodiment of a complete philosophy is an antagonism thoroughly wholesome. Once implanted in Mr. Herbert Spencer's mind, it is curious to observe how admirably it illustrates the idea of development. Having first sought some shelter of authority under words of the great prophet himself, he becomes more and more aggressive against the pretenders to his authority. His grumbles against them become loud and louder as he proceeds. He speaks of "those who have committed themselves to the current exclusive interpretation."[1] He observes upon "inattention and reluctant attention" as leading to the ignoring of facts. He speaks of "alienation from a belief" as "causing naturalists to slight the evidence which supports that belief, and refuse to occupy themselves in seeking further evidence." He compares their blindness now respecting the insufficiency of natural selection with the blindness of naturalists to the facts of evolution before Darwin's

[1] P. 581.

book appeared.   He marshals and reiterates the obvious considerations which
prove that the development of animal
forms must necessarily depend on an
immense number and variety of adjusted
changes in many different organs, all
co-operating with each other, and all
nicely adjusted to the improved functional actions in which they must all par-
take.   He reduces to a numerical com-
putation the practical impossibility of
such changes occurring as the result of
accident.   He tells his opponents that
the chances against any adequate re-
adjustments fortuitously arising "must
be infinity to one."[1]   But more than this :
he not only repels the Darwinian factor
as adequate by itself, but, advancing in
his conclusions, he declares that it must
be eliminated altogether.   On further
consideration he tells us that in his

[1] P. 571.

opinion it can have neither part nor lot
in this matter. He insists that the correlated changes are so numerous and so
remote that the greater part of them
cannot be ascribed (even) in any degree
to the mere selection of favourable variations.[1] Then facing the opponents
whose mingled credulities and increduli-
ties he has so offended, he rebukes their
fanaticisms according to a well-known
formula : " Nowadays," he says, "most
naturalists are more Darwinian than Mr.
Darwin himself."[2]

This is most true ; and Mr. Herbert
Spencer need not be the least sur-
prised. All this happens according
to a law. When a great man dies,
leaving behind him some new idea
—new either in itself or in the use he
makes of it—it is almost invariably seized
upon and ridden to the death by the

---

[1] P. 574.   [2] P. 584.

shouting multitudes who think they
follow him.   Mr. Herbert Spencer here
directs upon their confusions the search-
ing light of his analysis.   He most
truly distinguishes Darwin's hypothesis
in itself, first from the theory of "organic
evolution in general," and secondly from
"the theory of evolution at large."   This
analysis roughly corresponds with the
distinctions I have pointed out in the
preceding paper ; and when he points
to the confounding of these distinctions
under one phrase as the secret of wide
delusions, he has got hold of a clue by
which much further unravelling may be
done.   Guided by this clue, and in the
light of this analysis, he brings down
Darwin's theory to a place and a rank in
science which must be still further offen-
sive to those whom he designates as the
" mass of readers."   He speaks of it as
" a great contribution to the theory of

organic evolution." It is in his view a "contribution," and nothing more—a step in the investigation of a subject of enormous complexity and extent, but by no means a complete or satisfactory solution of even the most obvious difficulties presented by what we know of the structure and the history of organic forms.

It is no part of my object here to criticise in detail the value of that special conception with which Mr. Herbert Spencer now supplements the deficiencies of the Darwinian theory. He calls it "inheritance of functionally produced modifications," and he makes a tremendous claim on its behalf. He evidently thinks that it supplies not only a new and wholly separate factor, but that it goes a long way towards solving many of the difficulties of organic evolution. Nothing could indicate more strongly the immense proportions which

this idea has assumed in his mind than
the question which he propounds towards
the conclusion of his paper. Supposing
the new factor to be admitted, "do there
remain," he asks, "no classes of organic
phenomena unaccounted for?" Wonderful question, indeed! But at least it is
satisfactory to find that his reply is more
rational than his inquiry : "to this question, I think it must be replied that there
do remain classes of organic phenomena
unaccounted for. It may, I believe, be
shown that certain cardinal traits of
animals and plants at large are still un-
explained"; and so he proceeds to the
second paper, in which the still refractory
residuum is to be reduced.

Whatever other value may attach to
an attempt so ambitious, it is at least
attended with this advantage, that it
leads Mr. Herbert Spencer to follow up
the path of "further consideration" into

the phrases and formulæ of the Darwinian hypothesis. And he does so with memorable results. What he himself always aims at is to obliterate the separating lines between the organic and the inorganic, and to reduce all the phenomena of life to the terms of such purely physical agencies as the mechanical forces, — light, heat, and chemical affinity, etc. In this quest he finds the Darwinian phrases in his way. Accordingly, although himself the author and inventor of the most popular among them, he turns upon them a fire of most destructive criticism. He allows them to be, or to have been, "convenient and indeed needful"[1] in the conduct of discussion, but he condemns them as "liable to mislead us by veiling the actual agencies" in organic evolution. That very objection which has always

[1] P. 749.

been made against all phrases involving the idea of creation—that they are meta-phorical—is now unsparingly applied to Darwin's own phrase "natural selection." Its "implications" are pronounced to be "misleading." The analogies it points at are indeed definite enough, but unfortunately the "definiteness is of a wrong kind." "The tacitly implied 'nature' which selects is not an embodied agency analogous to the man who selects artificially." This objection cuts down to the very root of the famous formula, and to that very element in it which has most widely commended it to popular recognition and acceptance. But this is not all. Mr. Herbert Spencer goes, if possible, still deeper down, and digs up the last vestige of foundation for the vast but rambling edifice which has been erected on a phrase. The special boast of its worshippers has always been that it represented and embodied that great reform which removed the processes of organic evolution once and for ever from the dominion of deceptive metaphor, and founded them for the first time on true physical causation. But Mr. Herbert Spencer will have none of this. The whole of this pretension goes by the board. He pronounces upon it this most true and emphatic condemnation, "The words natural selection do not express a cause in the physical sense."[1] It is a mere "convenient figure of speech."[2]

But even this is not enough to satisfy Mr. Spencer in his destructive criticism. He goes himself into the confessional. He had done what he could to amend Darwin's phrase. He had "sought to present the phenomena in literal terms rather than metaphorical terms," and in

[1] P. 749.   [2] P. 750.

this search he was led to "survival of
the fittest." But he frankly admits that
" kindred objections may be urged
against the expression" to which this
leading led him. The first of these
words in a vague way, and the second
word in a clear way, calls up an idea
which he must admit to be "anthropocentric." What an embarrassment it is
that the human mind cannot wholly turn
the back upon itself. Self-evisceration,
the happy despatch of the Japanese, is
not impossible or even difficult, although
when it is done the man does not expect
to continue in life. But self-evisceration
by the intellectual faculties is a much
more arduous operation, especially when
we expect to go on thinking and defining as before. It is conceivable that a
man might live at least for a time without his viscera, but it is not conceivable
that a mind should reason with only

some bit or fragment of his brain. In the mysterious convolutions of that mysterious substance there are, as it were, a thousand retinæ—each set to receive its own special impressions from the external world. They are all needed; but they are not all of equal dignity. Some catch the lesser and others catch the higher lights of nature; some reflect mere numerical order or mechanical arrangement, whilst others are occupied with the causes and the reasons, and the purposes of these. Some philosophers make it their business to blindfold the facets which are sensitive to such higher things, and to open those only which are adapted to see the lower. And yet these very men generally admit that the faculties of vision which see the higher relations are peculiarly human. They are so identified with the human intellect that they can hardly be separated. And hence they are called anthropomorphic, or, as Mr. Spencer prefers to call them, "anthropocentric." This close association—this characteristic union—is the very thing which Mr. Spencer dislikes. Yet the earnest endeavours of Mr. Spencer to get out of himself—to eliminate every conception which is "anthropocentric"—have very naturally come to grief. "Survival"? Does not this word derive its meaning from our own conceptions of life and death? Away with it, then. What has a true philosopher to do with such conceptions? Why will they intrude their noxious presence into the purified ideas of a mind seeking to be freed from all anthropocentric contamination? And then that other word "fittest," does it not still more clearly belong to the rejected concepts? Does it not smell

of the analogies derived from the morti-
fied and discarded members of intelligence and of will ?   Does it not suggest
such notions as a key fitting a lock, or a
glove fitting a hand, and is it worthy of
the glorified vision we may enjoy of
Nature to think of her correlations as
having any analogy with adjustments
such as these ?   In the face of the
innumerable and complicated adjustments of a purely mechanical kind which
are conspicuous in organic life, Mr.
Spencer has the courage to declare that
"no approach" to this kind of fitness
"presentable to the senses" is to be
found in organisms which continue to
live in virtue of special conditions.

Where materials are so abundant it is
hard to specify.   But I am tempted to
ask whether Mr. Spencer has ever heard
of the ears, the teeth, above all the
finger of the Aye-aye, the wonderful

beast that lives in the forests of Mada-
gascar, and is very nicely fitted indeed
to prey upon certain larvæ which burrow
up the pith of certain trees? Here we
see examples of fitting in a sense as
purely mechanical as he could possibly
select from human mechanism. The
enormous ears are fitted to hear the
internal and smothered raspings of the
grub. The teeth are fitted for the work
of cutting-chisels, whilst one finger is
reduced to the dimension of a mere
probe, armed with a hooked claw to
extract the larvæ. The fitting of this
finger-probe into the pith-tube of the
forest bough is precisely like the fitting
of a finger into a glove. It is strange
indeed that Mr. Spencer should deny
the applicability of the word fitness, in
its strictest " glove " sense, to adaptations such as these. Yet he does
deny it in words emphatic and precise.

Neither the organic structures themselves—he proceeds to say—nor their
individual movements are related in any
analogous way to the things and actions
in the midst of which they live. Having
made this marvellous denial, he reiterates
in another form his great confession—
his *gran rifiuto*—that his own famous
phrase, although carefully designed to
express self-acting and automatic physical
operations, is, after all, a failure. And
this result he admits not only as proved,
but as obviously true. His confession
is a humble one. " Evidently," he says,
"the word fittest as thus used is a figure
of speech." [1]

This elaborate dissection and condemnation by Mr. Herbert Spencer of
both the two famous phrases which have
been so long established in the world
as expressing the Darwinian hypothesis

[1] P. 751.

—his emphatic rejection of the claim of either of them to represent true physical causation—his sentence upon both of them that they are mere figures of speech—is, in my judgment, a memorable fact. As regards Mr. Spencer himself, it is a creditable performance and an honourable admission. It is one of the high prerogatives of the human mind to be able to turn upon its own arguments, and its own imaginings, the great weapon of analysis. There are in all of us, not only two voices, but many voices, and splendid work is done when the higher faculties call upon the lower to give an account of what they have said and argued. Often and often, as the result of such a call, we should catch the accents of confession saying, "We have been shutting our eyes to the deepest truth, keeping them open only to others which were comparatively superficial.

We have been trying to conceal this by the invention of misleading phrases —full of loose analogies, of vague and deceptive generalities."

Most unfortunately, however, the special peculiarity of Mr. Spencer's introspection appears to be that it is the lower intellectual faculties which are calling the higher to account. The merit of Darwin's phrase lay in its elasticity—in its large elements of metaphor taken from the phenomena of mind. Mr. Spencer's phrase had been carefully framed, he tells us, to get rid of these. His great endeavour was to employ in the interpretation of nature only those faculties which see material things and the physical forces. Those other faculties which see the adjustments of these forces to purpose—to the building up of structures yet being imperfect, and to the discharge of functions yet lying in the future—it was his desire to exclude or silence. This was his aim, but he now sees that he has failed. In spite of him the higher intellectual perceptions have claimed admittance, and have actually entered. He now calls on the humbler faculties to challenge this intrusion, and to assert their exclusive right to occupy the field. The " survival of the fittest " had been constructed to be their fortress. But the very stones of which it is built—the very words by which the structure is composed—are themselves permeated with the insidious elements which they were intended to resist. The " survival of the fittest " is a mere redoubt open at the back, or a fort which can be entered at all points from an access underground. And so, like a skilful general, Mr. Spencer has ordered a complete evacuation of the works.

But in giving up this famous phrase

Mr. Spencer does not give up his purpose
—which, indeed, is one of the main
purposes of his philosophy—namely, to
build up sentences and wordy structures
which shall eliminate, as far as it is
possible to do so, all those aspects of
natural phenomena which are human,
that is to say, those aspects which reflect
at all an intellectual order analogous with,
or related to, our own. " I have elabor-
ated this criticism," he says, " with the
intention of emphasising the need for
studying the changes which have gone
on, and are ever going on, in organic
bodies from an exclusively physical point
of view." [1] And so, new formulæ are
constructed to explain and to illustrate
how this is to be done. " Survival "
suggesting the " human view " of life
and death must be dismissed. How,
then, are they to be described? They

[1] P. 751.

D

are "certain sets of phenomena." Their true physical character is "simply groups of changes." In thinking of a plant, for example, we must cease to speak of its living or dying. "We must exclude all the ideas associated with the words life or death."[1] What we do know, physically, is thus defined: "That there go on in the plant certain inter-dependent processes in presence of certain aiding or hindering influences outside of it; and that in some cases a difference of structure or a favourable set of circumstances allows these inter-dependent processes to go on for longer periods than in other cases."

How luminous! Milton spoke of his own blindness as "knowledge at one entrance quite shut out." But here we have a specimen of the verbal devices by which knowledge at all entrances may

[1] P. 751.

be carefully excluded. Life is certain "interdependent processes." Yes, certainly. But so is death. And so is everything else that we know of or can conceive. The words devised by Mr. Herbert Spencer to represent the "purely physical" view of life and death are words which present no view at all. They are simply a thick fog in which nothing can be seen. Except in virtue of this character of general opacity, they are wholly useless for Mr. Spencer's own purpose as well as for every other. He seeks to exclude mind. But he fails to do so. He seems to think that when he has found a collocation of words which do not expressly convey some particular idea, he has therein found words in which that idea is excluded. This is not so. Words may be so vague and abstract as to signify anything or nothing. If under the word "fitness" human ideas of adjustment and design are apt to insinuate themselves, assuredly the same ideas not only may, but must be comprehended under such a phrase as "interdependent processes." Painting, for example, is an interdependent process, and both in its execution and results its interdependence lies in purely physical combinations of visible and touchable materials. Yet Sir Thomas Lawrence spoke with literal truth when he snubbed a questioner as to the mechanics of his art by telling him that he mixed his colours with brains. The whole of chemical science consists in the knowledge of interdependent processes which are (what we call) purely physical, whilst the whole science of applied chemistry involves those other interdependent processes which involve the co-operation of the human mind and will.

We have, then, in this new phrase a
perfect specimen of one favourite method
of Mr. Herbert Spencer in his dealing
with such subjects; and the weapon of
analysis which he turns so successfully
against his own old phrase when he
wishes to abandon it, can be turned with
equal success not only against all sub-
stitutes for it, but against the whole
method of reasoning of which it was
an example. The verbal structures of
definition which abound in his writings
always remind me of certain cloud-forms
which may sometimes be seen in the
western sky, especially over horizons of
the sea. They are often most glorious
and imposing. Great lines of towers
and of far-reaching battlements give the
impression at moments of mountainous
solidity and strength. But as we gaze
upon them with wonder, and as we fix
upon them a closely attentive eye, the

edges are seen to be as unsteady as at first they appeared to be enduring. If we attempt to draw them we find that they melt into each other, and that not a single outline is steady for a second. In a few minutes whole masses which had filled the eye with their majesty, and with impressions as of the everlasting hills, dissolve themselves into vapour and melt away.

Such are the cloud-castles which mount upon the intellectual horizon as we scan it in the representations of the mechanical philosophy. Nothing can be more fallacious than the habit of building up definitions out of words so vague and abstract that they may signify any one of a dozen different things, and the whole plausibility of which consists in the ambiguity of their meanings. It is a habit too which finds exercise in the alternate amusement of wiping out of

words which have a definite and familiar
sense everything that constitutes their
force and power. Let us take for
example the word "function." There is
no word, perhaps, applicable to our
intellectual apprehensions of the organic
world, which is more full of meaning,
or of meaning which satisfies more
thoroughly the many faculties concerned
in the vision and description of its facts.
The very idea of an organ is that of an
apparatus for the doing of some definite
work, which is its function. For the
very reason of this richness and fulness
of meaning,—in this word conjoined with
great precision,—it is unfitted for use in
the vapoury cloud-castles of definition
which are the boasted fortresses of ideas
purely physical. And yet function is a
word which it is most difficult to dis-
pense with. The only alternative is to
reduce it to some definition which wipes

out all its special signification. Accordingly, Mr. Herbert Spencer has defined function as a word equivalent to the phrase "transformations of motion"[1]— a phrase perfectly vague, abstract, and equally applicable to function or to the destruction of it,—to the processes of death or the processes of life,—to the phenomena of heat, of light, or electricity, —and completely denuded of all the special meanings which respond to our perception of a whole class of special facts.

Of course the attempt breaks down completely to describe the facts of nature in words too vague for the purpose, or in words rendered sterile by artificial eliminations. It is not Darwin only, who had at least no dogma on this subject to bind him—it is Mr. Spencer himself who continually breaks down in the attempt, far more completely than he

[1] *Principles of Biology*, vol. i. p. 4.

now admits he failed in the "survival of the fittest." The human element involved or suggested in the idea of fitness is nothing to the anthropomorphism, or " anthropocentricity," of the expressions into which he slips, perhaps unawares, when he is face to face with those requisites of language which arise out of the facts of observation, and out of the necessities of thought. Thus in the midst of an elaborate attempt to explain in purely chemical and physical aspects the composition and attributes of protein, or protoplasm—assumed to be the fundamental substance of all organisms—he breaks out into the following sentence, charged with teleological phraseology : " So that while the composite atoms of which organic tissues are built up possess that low molecular mobility fitting them for plastic purposes, it results from the extreme molecular

mobilities of their ultimate constituents
that the waste products of vital activity
escape as fast as they are formed." [1]
Now, what is the value of sentences
such as this? As an explanation, or
anything approaching to an explanation,
of the wondrous alchemies of organic
life, and especially of the digestive processes—of the appropriation, assimilation, and elimination of external matter
—this sentence is poor and thin indeed.
But whatever strength it has is entirely
due to its recognition of the fact that not
only the organism as a whole, but the
very materials of which it is "built up,"
are all essentially adaptations which are
in the nature of "purposes," being indeed
contrivances of the most complicated
kinds for the discharge of functions of a
very special character.

What, then, is the great reform which

[1] *Principles of Biology*, vol. i. p. 24.

these new verbal forms are intended to effect in our conception of the factors in organic evolution? The popular and accepted idea of them has been largely founded on the language of Darwin and of Mr. Spencer himself. But that language has been deceptive. The needed reform consists in the more complete expulsion of every element that is "anthropocentric." In order to interpret Nature we must stand outside ourselves. The eye with which we look upon her phenomena must be cut off, as it were, from the brain behind it. The correspondences which we see, or think we see, between the system of things outside of us and that system of things inside of us which is the structure of our own intelligence, are to be discarded. This is the luminous conception of the new philosophy. Science has hitherto been conceived to be the reduction of

natural phenomena to an intelligible
order.   But the reformed idea is now to
be that our own intelligence is the one
abounding fountain of error and decep-
tion.   It is not merely to be disciplined
and corrected, but it is to be eliminated
altogether.   It is to be hounded off and
shouted down.

It is very clear what all this must end
in.   The demand made upon us in its
literal fulness is a demand impossible and
absurd.   We cannot stand outside ourselves.   We cannot look with eyes other
than our own.   We cannot think except
with the faculties of our own intellectual
nature.   It is impossible, and if it were
possible, it would be absurd.   We are
ourselves a part of nature—born in it,
and born of it.   The analogies which
the disciplined intellect sees in external
nature are therefore not presumably
false, but presumably true, or at the

least substantially representative of the truth.

But the new veto on anthropocentric thought, although helpless to expel it, is quite competent to cripple and degrade it. It cannot exclude our own faculties, but it may select and favour the lowest, the humblest, the most elementary, the most blunt, the least perceptive. It may silence the highest, the acutest, the most penetrating, the most intuitive, those most in harmony with the highest energies in the world around us. All this the new doctrine may do, and does.

Accordingly the very first instance given to us of the new philosophy is a striking illustration of its effects. It fixes the attention on mere outward and external things. It seeks for the first and best explanation of organic beings in the mere mechanical effects of their surroundings. The physical forces

which act upon them from outside—the
water or the air that bathes them—the
impacts of etherial undulations in the
form of light—the vibrations of matter in
contact with them in the form of heat—
these are conceived of as the agencies
principally concerned. The analogies
suggested are of the rudest kind. Old
cannon-balls rust in concentric flakes.
Rocks weather into such forms as rock-
ing-stones.[1] But the grand illustration
is taken from the pebbles of the Chesil
beach.[2] These are to introduce us to
the true physical conception of the
wonderful phenomena of organic life.
May not the unity of the vertebrate
skeleton, through an immense variety of
creatures, be typified by the roundness
and smoothness common to the stones
rolled along the southern beaches of
England from Devonshire to Weymouth?

[1] P. 755.	[2] P. 752.

The diversities of those creatures, again, however multitudinous in character, may they not all be pictured as analogous with the varying sizes into which water sifts and sorts the sizes of rolled stones?

But presently we see in another form the work of "natural selection" by a mind deliberately divesting itself of its own higher faculties, and choosing in consequence to exert only those which are simple and almost infantile. The question naturally arises—what is the most universal peculiarity and distinction of organic forms? When we get rid of ourselves,—when we stand outside of our own anthropocentric position and consult only the faculties which are most purely physical,—we shall be compelled to reply that the great speciality of organic forms is the "differentiation of their outside from their inside."[1]   They have all an

[1] P. 755.

outside and an inside, and these are different. They begin with a cell, and a cell is a blob of jelly with a pellicle or thin membrane on the outside. Do we not see in this the mechanical action of the surrounding medium? The membrane may come from a chill on the outside, or the pressure of the medium. Does not a little oil form itself into a sphere in water, or a little water into a drop in air? And so from one step to another, cannot we conceive how particles of protein become cells, and how one cell gets stuck to another, and the groups to groups—all with insides and outsides "differentiated" from each other, and so they can all be pressed and compacted and squeezed together until the organism is completed?[1]

Such or such like are the images presented to enable us to conceive the

[1] Pp. 756-758.

purely physical view of the beginnings of life. Their own genesis is obvious. It is true that all or nearly all organisms have a skin. Most if not all of them begin, so far as seen by us, in a nucleated cell. The external wall of these cells is often a mere pellicle. It is true also that one essential idea of life is separation or segregation from all other things. This is an essential part of our ideas of individuality and of personality. If a pellicle or skin round a bit of protein be taken as the symbol of all that is involved in this idea of life, then "outness" and "inness" may be tolerated as a very rude image of one of the great peculiarities of all organic life. It may even be regarded as a symbol of the thoughts expressed in the solemn lines—

> Eternal form shall still divide
> The eternal soul from all beside.

E

But if "outer" and "inner" are used to express the idea of some essential mechanical separation between different parts of the same organism, so that one part may be represented as more the result of surrounding forces than another —then this rude and mechanical illustration is not only empty, but profoundly erroneous. The forces which work in and upon organic life know nothing of outness and inness. They shine through the materials which they build up and mould, as light shines through the clearest glass. Even the most purely physical of those concerned are independent of such relations. Gravitation knows nothing of inness and outness. The very air, which seems so external to us, does not merely bathe or lave the skin, but permeates the blood, and its elements are the very breath of life in every tissue of the body. The more secret forces of vitality deal at their will with outness and inness. The external surfaces of one stage are folded in and become most secret recesses at another. Organs which are outside in one animal, and are conspicuously flourished in the face of day with exquisite ornament of colour and of structure,[1] are in another animal hid away and carefully covered up. Nay, there are many cases in which all these changes are conducted in the same animal at different periods of life, and during conscious and unconscious intervals the whole creature is re-formed to fit it for new surroundings, for new media, and with new apparatuses adapted to them.

If Mr. Spencer wishes to cast any fresh light upon those factors of organic evolution respecting which he now confesses that Darwin's language and his

[1] As in the nudibranchiate mollusca.

own have been alike defective, he must fix our attention on something deeper than the differences between every organism and its own skin. His selection of this most superficial kind of difference as the first to dwell upon, is not merely wanting—it is erroneous. It hides and leads us off the scent of another kind of outsidedness and insidedness which is really and truly fundamental; namely, the insidedness, the self-containedness, of every organism as a whole with reference to all external forces. Nobody has pointed this out more clearly in former years than Mr. Spencer himself. The grand distinction between the organic and the inorganic lies in this—that the organic is not passive under the touch or impact of external force, but responds, if it responds at all, with the play of counter-forces which are essentially its own.

Organic bodies are not simply moved. They move themselves. They have "self-mobility."[1] They are so constituted that even when an external force acts as an excitement or a stimulus, the organic forces which emerge and act are much more complex and important—so much so that as compared with the results produced by these organic forces the direct results of the incident forces are "quite obscured."[2] Mr. Spencer even confesses that these two kinds of action are so different in their own nature that in strictness they "should not be dealt with together." But he adds that "the impossibility of separating them compels us to disregard the distinction between them." This is a most lame excuse for the careless—and a still worse excuse for the studied—use

[1] P. 757.
[2] *Principles of Biology*, vol. i. p. 43.

of ambiguous language which confounds the deepest distinctions in nature. It cannot be admitted. All reasonings on nature would be hopeless unless we could separate in thought many things which are always conjoined in action ; and this excuse is all the more to be rejected when the alleged impossibility of separation is used to cover an almost exclusive stress upon that one of the two kinds of action which is confessedly by far the feeblest, and of least account in the resulting work.

It seems to me, further, that there is another fatal fault in this attempt of Mr. Spencer to reform the language, and clear up the ideas of biological science. Besides the method of habitually using words so abstract as to be of necessity ambiguous—besides the further method of habitually expelling from definite words the only senses which give them value—Mr. Spencer often resorts, and does so conspicuously in this paper, to the scholastic plan of laying down purely verbal propositions and then arguing deductively from them as if they represented axiomatic truth. By the schoolmen this method was often legitimately applied to subjects which in their own nature admitted of its use, because those subjects were not physical but purely moral or religious, and in which consequently much depended on the clear expression of admitted principles of abstract truth. I will not venture to say that such verbal propositions embodying abstract ideas can have absolutely no place in physical science. We know as a matter of fact that they have led some great men to the first conception of a good many physical truths; and it is a curious fact that Dr. Joule, who in our own day has been the first to establish

the idea of the doctrine of the Conserva-
tion of Energy by proving through
rigorous experiment the mechanical
equivalent of heat, has said that "we
might reason *a priori* that the absolute
destruction of living force cannot possibly
take place because it is manifestly absurd
to suppose that the powers with which
God has endowed matter can be de-
stroyed, any more than they can be
created, by man's agency."[1]

Believing as I do in the inseparable
unity which binds us to all the verities
of nature, I should be the last to pro-
scribe the careful use of our own abstract
conceptions. But it is quite certain and
is now universally admitted that the
methods of Thomas Aquinas in his
Summa are full of danger when they are

[1] In a lecture delivered at Manchester, April 28,
1847. See *Strictures on the Sermon, etc.,* by B. St.
J. B. Joule, J.P., a pamphlet published 1887 (J.
Heywood, Manchester).

used in physical investigation. Yet as regards at least the tone of dogma and authority, and also as regards the method of reasoning, we have from Mr. Spencer in this paper the following wonderful specimen of scholastic teaching on the profoundest questions of organic structure : " At first protoplasm *could have* no proclivities to one or other arrangement of parts ; unless indeed a purely mechanical proclivity towards a spherical form when suspended in a liquid. At the outset it *must have been* passive. In respect of its passivity, primitive organic matter *must have been* like inorganic matter. No such thing as spontaneous variation *could have* occurred in it ; for variation implies some habitual course of change from which it is a divergence, and is therefore excluded where there is no habitual course of change." What possible knowledge can Mr. Spencer

possess of " primitive organic matter " ?
What possible grounds can he have for
assertions as to what it *must have been*,
and what it *must have done?* Surely
this is scholasticism with a vengeance.
Its words, its assumptions, and its claims
of logical necessity are all equally hazy,
inconclusive, and absolutely antagonistic
to the spirit of true physical science.

There is a passing sentence in one of
Darwin's works[1] which will often recur
to the memory of those who have
observed it. Speaking of the teleological or theological methods of describing nature, he says that these can
be made to explain anything. At first
sight this may seem a strange objection
to any intelligible method — that it is
too widely applicable. But Darwin's
meaning is in its own sphere as true as

[1] I have mislaid the reference, and quote from
memory.

it is important.   An explanation which
is good for everything in general, is good
for nothing in particular.   Explanations
which are indiscriminate can hardly be
also special and distinguishing.   In their
very generality they may be true, but
the truth must be as general as the terms
in which it is expressed.   Thus the
common phrase which we are in the
habit of applying to the wonderful adaptations of organic life when we call them
"provisions of nature" is a phrase of this
kind.   It satisfies certain faculties of the
mind, and these the highest, but it
affords no satisfaction at all to those
other faculties which ask not why but
how these adaptations are affected.   It
is an explanation applicable to all adaptations equally, and to no one of them
specially.   It takes no notice whatever
of the question, How?   It does not
concern itself at all with physical causes.

Darwin saw this clearly of such methods of explanation. But he did not see that precisely the same objection lies against his own. The great group of ideas metaphorically involved in his phrase of natural selection, and not successfully eliminated in the summary of it—survival of the fittest—is a group of the widest generality. It may be used to account for anything. The successful application of it to any organic adaptation, however special and peculiar, is so easy as to become a mere trick. We have only to assume the introduction of some primordial organisms—one or more —already formed with all the special powers and functions of organic life ; we have only to assume the inscrutable action of heredity ; we have only to assume, further, that it originates difference as well as transmits likeness ; we have only to assume, still further, that the variations so originated are almost infinite in variety, and that some of them are almost sure, at some time or another, to "turn up trumps," or in other words to be accidentally in a useful direction ; we have only to assume, again, that these will be somehow continued and developed through embryotic stages until they are fit for service ; we have only to assume, again, that there are adjustments by which serviceability, when transmuted into actual use, has power still further to improve all adaptations by some process of self-edification ; then, making all these assumptions, we may explain anything and everything in the organic world. But in such a series of assumptions we do not speak the language of true physical causation. This is what Mr. Spencer now confesses. "Natural selection," he says, "could operate

only under subjection."[1]   This is a
prolific truth.   It might have been discovered sooner.   Natural selection could
only select among things prepared for
and presented to its choice.   How—
from what physical causes—did these
come ?   Mr. Spencer's reply is, historically speaking, retrograde.   He goes back
to Lamarck, he reverts to "use and disuse," to "environment"—to surroundings—to the "medium and its contents."[2]
These again are mere phrases to cover
the nakedness of our own ignorance.
But I for one am thankful for the conclusion arrived at by a mind so acute
and so analytical as that of Mr. Spencer,
that "among biologists the beliefs
concerning the origin of species have
assumed too much the character of a
creed, and that while becoming settled
they have been narrowed.   So far from

[1] P. 768.   [2] *Ibid.*

further broadening that broader view
which Darwin reached as he grew older,
his followers appear to have retrograded
towards a more restricted view than he
ever expressed." The evil must have
gone far indeed when this great apostle
of Evolution has to plead so laboriously
and so humbly "that it is yet far too
soon to close the inquiry concerning the
causes of organic evolution." Too soon
indeed! That such an assumption
should have been possible, and that it is
virtually made, is part of the Great
Confession to which I have desired to
direct attention. I hope it will tend to
redeem the work of the greatest natural
observer who has ever lived from the
great misuse which has been often made
of it. There is no real disparagement
of that work in saying that the phrase
which embalmed it is metaphorical.
The very highest truths are conveyed in

metaphor. The confession of Mr. Spencer is fatal only to claims which never ought to have been made. Natural selection represents no physical causation whatever except that connected with heredity. Physically it explains the origin of nothing. But the metaphorical elements which Mr. Spencer wishes to eliminate are of the highest value. They refer us directly to those supreme causes to which the physical forces are "under subjection." They express in some small degree that inexhaustible wealth of primordial inception, of subsequent development, and of continuous adjustment, upon which alone selection can begin to operate. These are the supreme facts in nature. When this is clearly seen and thoroughly understood, Darwin's researches and speculations will no longer act as a barrier to further inquiry, as Mr. Spencer

complains they now do.   They will,
on the contrary, be the most powerful
stimulus to deeper inquiry, and to more
healthy reasoning.

CHAPTER II

MR. HERBERT SPENCER contributed to *The Nineteenth Century* in November 1895 an article entitled "Lord Salisbury on Evolution." The occasion of it arose out of the brief but significant comments on the Darwinian theory which formed part of Lord Salisbury's Presidential Address to the British Association at Oxford in 1894. In so far as that article is merely a reply to Lord Salisbury, it does not concern us here. But, like everything from Mr. Spencer's pen, it is full of highly instructive matter on the whole subject

to which it relates.   It takes a much
larger view of the problems of Biology
than is generally taken, and it deals
with them by a method which is excellent, so far as he carries it, and which
we can all take up and follow farther
than the point at which he stops.   That
method is to insist on a clear definition
of the words and phrases used in our
biological data and speculations.   No
method could be more admirable than
this.   It is one for which I have myself
a great predilection, and have continually used in all difficult subjects of
inquiry.   Such, pre-eminently, are the
problems presented by the nature and
history of organic life.   I propose, therefore, in these pages to accept Mr.
Spencer's method, and to examine what
light can come from it on this most
intricate of all subjects.

The leading idea of Mr. Spencer's

article is to assert and insist upon a
wide distinction between the "natural
selection" theory of Darwin and the
general theory of what Mr. Spencer
calls "organic evolution." He insists
and reiterates that even if Darwin's
special theory of natural selection were
disproved and abandoned, the more
general doctrine of organic evolution
would remain unshaken. I entirely
agree in this discrimination between
two quite separate conceptions. But I
must demand a farther advance on the
same lines — an advance which Mr.
Spencer has not made, and which does
not appear to have occurred to him as
required. Not only is Darwin's special
theory of natural selection quite separable from the more general theory of
organic evolution, but also Mr. Spencer's
own special version and understanding
of organic evolution is quite separable

from the general doctrine of development, with which, nevertheless, it is habitually confounded. It is quite as true that even if Mr. Spencer's theory of organic evolution were disproved and abandoned, the general doctrine of development would remain unshaken, as it is true that organic evolution would survive the demolition of the Darwinian theory of Natural Selection.

The great importance of these discriminations lies in this—that both the narrow theory of Darwin, and also the wider idea of organic evolution, have derived an adventitious strength and popularity from elements of conception which are not their own—elements of conception, that is to say, which are not peculiar to them, but common to them and to a much larger idea—a far wider doctrine—which has a much more indisputable place and rank in the facts of nature, and in the universal recognition of the human mind.

Let us, therefore, unravel this entanglement of separable ideas much more completely than Mr. Spencer has done. And for this purpose let us begin at the bottom—with the one fundamental conception which underlies all the theories and speculations that litter the ground before us. That conception is simply represented by the old familiar word, and the old familiar idea—of growth or development. It is the conception of the whole world, in us and around us, being a world full of changes, which to-day leave nothing exactly as it was yesterday, and which will not allow to-morrow to be exactly as to-day. It is the conception of some things always coming to be, and of other things always ceasing to be—in endless sequences of cause and of effect. It has this great advantage—that it is not a mere doctrine or a mere theory, nor an hypothesis, but a visible and undoubted fact. Nobody can deny or dispute it. Nowhere has it been more profoundly expressed and described, in its deepest meanings and significance, than in the words of that great metaphysician— whoever he may be—who wrote the Epistle to the Hebrews, when he describes the Universe as a system in which "the things which we see were not made of things that do appear." That is to say, that all its phenomena are due to causes which lie behind them, and which belong to the Invisible. Nor can we even conceive of its being otherwise. The causes of things— whatever these may be—are the sources out of which all things come, or are developed. What these causes are has been the Great Quest, and the great incentive to inquiry, since human thought began. But there never has been any doubt, or any failure, on the part of man to grasp the universal fact that there is a natural sequence among all things, leading from what has been to what is, and to what is to be Whether he could apprehend or not the processes out of which these changes arise, he has always recognised the existence of such processes as a fact.

One might almost suppose from much of the talk we have had during the last thirty years about development, that nobody had ever known or dwelt upon this universal fact until Lamarck and Darwin had discovered it. But this is far from being true. The recognition of the fact has been an element in all philosophies since philosophy began. All the new theories, and, indeed, all possible theories which may supplant or supplement them, are nothing but guesses at the details of the processes through which causation has long been recognised as working its way from innumerable small beginnings to innumerable great and complicated results. Every one of these guesses may be wrong in whole, or in essential parts, but the universal facts of growth and development in Nature remain as certain and as obvious as before.

It is a bad thing, at least for a time, when the undoubtedness of a great general conception such as this—of the continuity of causation and of the gradual accumulation of its effects— gets hooked on (as it were) in the minds of theorists to their own little fragmentary fancies as to particular modes of operation. But it is a worse thing still when this spurious and accidental affiliation becomes so established in the popular mind that men are afraid not to

accept the fancies lest they should be
thought to impugn the facts or to deny
admitted and authoritative truths.   Yet
this is exactly what has happened with
the Darwinian theory.   The very word
"development" was captured by the
Darwinian school as if it belonged to
them alone, and the old familiar idea was
identified with theories with which it
had no necessary connection whatever.
Development is nowhere more conspicuous than in the history of human
inventions; the gun, the watch, the steam-
engine, and our new electric machines
have all passed through many stages
of development, every step in which
is historically known.   So it is with
human social and political institutions,
when they are at all advanced.   But
this kind and conception of development has nothing whatever to do with
the purely physical conceptions involved

in the Darwinian theory. The idea, for example, of one suggestion arising out of another in the constructive mind of man, is a kind of development absolutely different from the idea of one specific kind of organic structure being born by ordinary and physical generation of quite different parents without the directing agency of any mind at all. Our full persuasion of the perfect continuity of causation does not compel us to accept, even for a moment, the idea of any particular cause which may be obviously incompetent, far less such as may be conspicuously fantastic. Nor—and this is often forgotten—does the most perfect continuity of causes involve, as a necessary consequence, any similar continuity in their visible effects. These effects may be sudden and violent, although the previous working has been slow and even infinitesimally gradual. In short, the

general idea of development is a conception which remains untouched whether
we believe, or do not believe, in any particular hypotheses which may profess to
explain its steps.

Mr. Spencer, then, adopts an excellent method when he insists upon discriminations such as these between
very different things jumbled together
and concealed under loose popular
phrases.   But, unfortunately, he fails
to pursue this method far enough.
There is great need of the farther
application of it to his own language.
He tells us that Darwinism is to be
carefully distinguished from what he
calls "organic evolution."   Darwinism
he defines in the phrases of its author.
But organic evolution he does not
define so as to bring out the special
sense in which he himself always uses
it.   On the contrary, he employs words

to define organic evolution which systematically confound it with the general idea of development, whilst concealing this confusion under a change of name. The substitution of the word " evolution " for the simpler word " development " has, in this point of view, an unmistakable significance.   I do not know of any real difference between the two words, except that the word " development " is older and more familiar, whilst " evolution " is more modern, and has been more completely captured and appropriated by a particular school.   But Darwin's theory is quite as distinctly and as definitely a theory of organic evolution as the theory of which Mr. Spencer boasts that it will remain secure even if Darwinism should be abandoned.   Both these theories are equally hypotheses as to the particular processes through which development has held its way in that department of

Nature which we know as organic life. But it is quite possible to hold, and even to be certain, that development has taken place in organic forms, without accepting either Darwin's or Mr. Spencer's explanation of the process. They both rest—as we shall see—upon one and the same fundamental assumption ; and they are both open to one and the same fundamental objection— viz. the incompetence of them both to account for, or to explain, all the facts, or more even than a fraction of the facts, with which they profess to deal.

In order to make this plain we have only to look closely to the peculiarities of the Darwinian theory, and ascertain exactly how much of it, or how little of it, is common to the theory which Mr. Spencer distinguishes by the more general title of organic evolution. Darwin's theory can be put into a few very

simple propositions—such as these : All
organisms have offspring. These offspring have an innate and universal
tendency to variation from the parent
form. These variations are indeter-
minate—taking place in all directions.
Among the offspring thus varying, and
between them and other contemporary
organisms, there is a perpetual competi-
tion and struggle for existence. The
variations which happen to be advantageous in this struggle—from some
accidental better fitting into surrounding
conditions—will have the benefit of that
advantage in the struggle. They will con-
quer and prevail; whilst other variations,
less advantageous, will be shouldered out
—will die and disappear. Thus step by
step, Darwin imagined, more and more
advantageous varieties would be accidentally but continually produced, and
would be perpetuated by hereditary

transmission. By this process, pro-
longed through ages of unknown duration, he thought it was possible to
account for the origin of the millions
of different specific forms which now
constitute the organic world. For this
theory, as we all know, Darwin adopted
the phrase Natural Selection. It was
an admirable phrase for giving a certain
plausibility and vogue to a theory full
of weaknesses not readily detected. It
spread over the confused and disjointed
bones of a loose conception the ample
folds of a metaphor taken from wholly
different and even alien spheres of experience and of thought. It resorted
to the old, old Lucretian expedient of
personifying Nature, and lending the
glamour of that Personification to the
agency of bare mechanical necessity,
and to the coincidences of mere fortuity.

Selection means choice by a living

agent out of some pre-existing things. The skilful breeders of doves and dogs and horses were, in this phrase, taken as the type of Nature in her production and in her guidance of varieties in organic structure. Darwin did not consciously choose this phrase because of these tacit implications of Mind and Will. He was in all ways simple and sincere, and he no more meant to impose upon others than on himself when he likened the operations of Nature in producing new species to the foreseeing skill of the breeder in producing new and more excellent varieties in domestic animals. Nevertheless, as a fact, this implication is indelible in the phrase, and has always lent to it more than half its strength, and all its plausibility. Darwin was led to it by an intellectual instinct which is insuperable—viz. the instinct which sees the

highest explanations of Nature in the
analogies of mental purpose and direction. The choice by Darwin of the
phrase Natural Selection was in itself an
excellent example of its only legitimate
meaning. He did not invent either
the idea or the phrase of Selection.
He found it existing and familiar. He
took it from the literature of the farm-
yard, of the kennel, and of the stable.
He told Lyell that it was constantly
used in all books of breeding. It was
his own intellectual nature that made the
choice, selecting it out of old materials.
These materials were gathered out of
the experience of human life, and out
of the nearest analogies of that natural
system of which Man is the highest
visible exponent. But Darwin neither
saw nor admitted its implications. The
great bulk of his admirers have not been
exactly in the same condition of mind, for

they have rejoiced in his theory for the very reason that it rested mainly on the idea of fortuity, or of mechanical necessity, and excluded altogether the competing idea of mental direction and design. In this they were more Darwinian than Darwin himself. He assumed, indeed, that variations were promiscuous and accidental; but he did so avowedly only because he did not know any law directing and governing their occurrence. His fanatical followers went farther. They have assumed that on this question there is nothing to be known, and that the rule of accident and of mechanical necessity had for ever excluded the agency of Mind.

Let us now ask of ourselves the question, Which of those two elements in Darwin's theory—the element of accident and of mechanical necessity, or the element of a directing agency in the

path of variation—has better stood the
test of thirty years' discussion, and
thirty years of closer observation ?   Can
there be any doubt on this ?   Year
after year, and decade after decade,
have passed away, and as the reign of
terror which is always established for
a time to protect opinions which have
become a fashion, has gradually abated,
it has become more and more clear that
mere accidental variations, and the mere
accidental fitting of these into external
conditions, can never account for the
definite progress of correlated adjustments and of elaborate adaptations,
along certain lines, which are the most
prominent of all the characteristics of
organic development.   It would be as
rational to account for the poem of the
Iliad, or for the play of *Hamlet*, by
supposing that the words and letters
were adjusted to the conceptions by

II   SELECTION CAN'T ORIGINATE   85

some process of "natural selection" as to account, by the same formula, for the intricate and glorious harmonies between structure and functions in organic life.

It has been seen, moreover, more and more clearly, that whilst that branch of his theory which rested on fortuity was obviously incompetent, that other branch of it which claimed affiliation with the directing agency of mind and choice was as incompetent as its strange ally. Selection, as we know it, cannot make things; it can only choose among materials already made and open to the exercise of choice. Therefore selection, whether by man or by what men are pleased to call Nature, can never account for the origin of anything. Then, other flaws, equally damaging to the theory, have been, one after another, detected and exposed. There are a multitude of structures in which no utility can be detected, but in which, nevertheless, development has certainly held its way, steadily and often with marvellous results. Nor is it less certain that there are some characteristics of many organisms which can be of no use whatever to themselves, but are of immense use to other organisms which find them nutritive and delicious to devour or valuable to domesticate and enslave. In short, men have been more and more coming to perceive that, as Agassiz once wrote to me in a private letter, " the phenomena of organic life have all the wealth and intricacy of the highest mental manifestations, and none of the simplicity of purely mechanical laws."

What, then, is Mr. Spencer's own verdict on the Darwinian theory of Natural Selection? He confesses at once that it gives no explanation of some of the phenomena of organic life. But he specifies one example which makes us doubt whether in his mouth the admission is of any value. The effects of use and disuse on organs are, he says, not accounted for.[1] The example is surely a bad one as any measure, or even as any indication, of the quality and variety of biological facts which altogether outrun the ken of Darwinism. In my opinion, it is no example at all — because the phrase Natural Selection is so vague and metaphorical in its implications that it may be made to cover and include quite as good an explanation of the effects of disuse as of a thousand other familiar facts. Organs, when fit and ready for use, are strengthened by healthy exercise. Organs, on the other hand, of the same kind, are weakened and

[1] P. 740.

atrophied by long - continued disuse. This is a familiar fact. What can be more easy than to translate this general fact into Darwinese phraseology ? Nature has a special favour for organs put to use. She strengthens them more and more by a process falling well under the idea of Natural Selection. In like manner, Nature deals unfavourably with organs which are allowed to be idle and inactive. She places them at a disadvantage, and they tend to perish.

The truth is, that the phrase Natural Selection and the group of ideas which hide under it is so elastic that there is nothing in heaven or on earth that by a little ingenuity may not be brought under its pretended explanation. Darwin in 1859-60 wondered " how variously " his phrase had been " misunderstood." The explanation is simple : it was because of those vague and loose analogies which

are so often captivating.   It is the same
now, after thirty-six years of copious
argument and exposition.   Darwin ridi-
culed the idea which some entertained
that Natural Selection "was set up as
an active power or deity"; yet this is
the very conception of it which is at this
moment set up by one of the most faith-
ful worshippers in the Darwinian Cult.
Professor Poulton of Oxford gives to
Natural Selection the title of "a motive
power" first discovered by Darwin.
This development is perfectly intelligible.
Nature is the old traditional refuge for
all who will not see the work of creative
mind.   Everything that is—everything
that happens—is and happens naturally.
Nature personified does, and is, our all
in all.   She is the universal agent, and
at the same time the universal product.
What she does she may easily be conceived as choosing to do, or selecting to

be done, out of countless alternatives
before her.   Then we have only to shut
our eyes, blindly or conveniently, to the
absolute difference between the idea of
merely selecting out of already existing
things, and of selecting by prevision out of
conceivable things yet to be—we have
only to cherish or even to tolerate this
gross confusion of thought—and then we
can cram into our theories of Natural
Selection the very highest exercises of
Mind and Will.   Let us carry out consistently the analogy of thought involved
in the agency of a human breeder; let
us emancipate this conception from the
narrow limits of operation within which
we know it to be humanly confined; let
us conceive a strictly homologous agency
in Nature which has power not merely to
select among organs already so developed
as to be fit for use, but to select and
direct beforehand the development of

organs through many embryotic stages of existence during which no use is possible; let us conceive, in short, an agency in Nature which keeps, as it were, a book in which "all our members are written, which in continuance are fashioned, when as yet there are none of them,"[1] then the phrase and the theory of Natural Selection may be accepted as at least something of an approach to an explanation of the wonderful facts of biological development.

But this is precisely the aspect of the Darwinian theory which Mr. Spencer dislikes the most. It is the aspect most adverse to his own philosophy. And as "natural rejection" is a necessary correlative of all conceptions of Natural Selection, so Mr. Spencer's intellectual instincts perceive this necessary antagonism, and lead him to dissent from

[1] Ps. cxxxix. 16.

Darwin's theory on account of that very element on which much of its popular success has undoubtedly depended. Mr. Spencer dismisses with something like contempt the ideas connected with the agency of a human breeder. He has, therefore, always condemned the phrase under which this idea is implied. He will have nothing to do with the conception of mind guiding and directing the course of development. Therefore, he has long suggested the adoption of an alternative phrase for the Darwin theory, which phrase is the "survival of the fittest." It has always seemed to me that the insuperable objection to this phrase is that it means nothing but a mere truism. If we eliminate from Darwin's theory the mental element of selection, and if we eliminate also, as we must do, the element of pure chance, which, of course, is nothing but a confession of ignorance, what is there remaining ?   Mr. Spencer's answer to this question is that the " survival of the fittest " remains.   Yes, but this is a mere restatement of certain facts under an altered form of words which pretends to explain them, whilst in reality it contains no explanatory element whatever.   The survival of the fittest?   Fittest for what?   For surviving.   So that the phrase means no more than this, that the survivor does survive.   It surely did not need the united exertions of the greatest natural observer of modern times and the reasonings of one of the most ingenious of modern philosophers to assure us of the truth of this identical proposition. Yet, in the article now under review, it is at least a comfort to find that Mr. Spencer confesses to the empty certitude which his phrase contains.   He says it is a self-evident proposition like an axiom in mathematics.[1]   The negation of it, he says, is inconceivable.   But if so, it tells us nothing.   If we do enter at all on the field of speculation on the origin and development of organic things, we do not need to be assured that the fittest things for surviving do, accordingly, and necessarily, survive.   What we want to know—or at least to have some glimpse of—is the processes of development, through which fitness has been attained for creatures moving along innumerable divergent paths of energy and of enjoyment.   A theory which, in answer to our inquiries on this high theme, tells us confessedly nothing but the self-evident proposition that the creatures fittest to survive do actually survive, is manifestly nothing but a mockery and a snare.

But Mr. Spencer has a substitute for

[1] Pp. 748, 749.

the Darwinian theory thus reduced to emptiness—something which, he says, lies behind and above it, and which only emerges with all the greater certainty when the ruins of that theory have been cleared away. This substitute is the generalised term "organic evolution." But what is this? Is it anything more than the general idea of development in its special application to organic life? No, it is nothing more. It is again the mere assertion of a self-evident proposition—that organic forms have been developed—somehow. We know it in the case of our own bodies and in the case of all contemporary living things. Mr. Spencer gives us no short and clear definition of what he means by organic evolution either in itself or as distinguished from the form of it taken in the Darwinian theory of natural selection. He refers to some of the characteristic features of all development, which are really sufficiently well known to all of us. Nothing that we see, or know, nothing that we can even conceive, is produced at once as a finished article, ready-made without any previous processes of growth. All this is no theory. It is a fact. Mr. Spencer laboriously counts up four or five great heads of evidence upon this subject, as if any one does or could dispute it. First comes Geology, with its long record of organic forms, showing, despite many gaps and breaks, on the whole an orderly procession from the more simple to the most complex structures. Secondly comes the science of Classification, the whole principle of which is founded on the possibility of arranging animal forms according to definite likenesses and affinities in structure. Thirdly comes the distribution of species — showing special likenesses

between the living fauna and the extinct
fauna of the great continents and islands
of the globe, which are most widely
separate from others, and suggesting
that, as the likeness has been continuous,
so it must be due to local continuities of
growth.   Fourthly there are the wonderful facts of Embryology, which are full
of suggestions to a like effect.   Then
there is another head of evidence, making
a fifth, which Mr. Spencer is disposed to
add to the other four—a head of evidence
which I venture to regard as even more
interesting and significant than any other
—that, namely, which rests on the occurrence of what are called Rudimentary
Organs in many animal frames—that is
to say, organs, or bits of structure, which,
in those particular creatures, are almost
or entirely devoid of any functional use,
but which correspond, more or less, with
similar organs in other animals where

H

they are in full, and all-important,
functional activity.

I accept all these five lines of evidence
as each and all confirmatory of the
leading idea of development—an idea
which I hold to be indisputably applicable to everything, and especially to
organic life. But Mr. Spencer is dream-
ing if he assumes that any, or all, of
these evidences prove either that particular theory of evolution which was
Darwin's, or that modification of it
which is his own. He seems to think,
and indeed expressly assumes, that the
only alternative to that theory is what he
calls the theory of " Special Creation."
But I do not know of any human being
who holds that theory in the sense in
which Mr. Spencer understands it. He
deals with what he calls Special Creation
very much as the late Professor Huxley
used to deal with the idea of a Deluge.

That is to say, he puts that idea into an absurd form, and then ascribes that absurdity to his opponents. Huxley used to picture a deluge as involving the idea of a mass of water, thousands of feet deep, holding its place at one time and over the whole globe, in defiance of the laws of gravitation, and especially of hydrostatics. It is a pity that Huxley did not live to see the venerable Sir Joseph Prestwich—the greatest authority on quaternary geology—avow his conviction that during that period of the earth's history there is a clear geological evidence that there must have been— at least over the whole Northern Hemisphere—some great submergence which was very wide, sudden, transitory, and extensively destructive to terrestrial life.

In like manner Mr. Spencer insists that those who have believed in Special Creation must believe that the bodies of

all animals appeared suddenly, ready-
made, complete in all their parts, out of
the dust of the ground and the elements
of the atmosphere. This, indeed, may
have been the crude idea of many men in
former times, in so far (which was very
little) as they gave themselves any time
to think, or to form any definite conceptions, on the meaning of the words they
used.   But the late Mr. Aubrey Moore,
in an interesting essay,[1] has reminded us
that it was the extravagant literalism of
Puritan theology which first embodied
in popular form this coarser view of
Creation, in a famous passage of *Para-
dise Lost.*[2] Yet this is a passage which
probably no man can now read, notwith-
standing the splendid diction of the poet,
without feeling the picture it presents to

[1] *Science and Faith*, 1889, "Darwinism and the
Christian Faith."
[2] Book vii.

be childish and grotesque. Mr. Moore
has reminded us, too, that both among
the Fathers and the Schoolmen of the
Christian Church there was no antipathy
to the idea that animals were, somehow,
genetically related to each other. I
doubt whether there is now any man
of common education who believes, for
example, that each of the many kinds of
wild pigeons which are spread over the
globe, and which are all so closely related
to each other by conspicuous similarities
of form, were all separately and individu-
ally created out of the raw materials of
nature.

Lord Salisbury in his Address says
that one thing Darwin has done has been
to destroy the doctrine of the immutability of species. This may be true of
absolute immutability, which can be
asserted of nothing that exists in this
world. Yet it does not follow that the

converse is true, namely, what may be
called the fluidity, or perpetual instability,
of species. There is at least one possible,
and even probable, alternative between
these two extreme alternatives. It is
surely a curious fact that the two
greatest naturalists of the modern world,
Cuvier and Linnæus, whose minds were
brought by their special pursuits into the
closest possible contact with the only
facts in Nature that have a direct bear-
ing on this question, were both of them
not only convinced of the stability of
species, but recognised it as the essential
foundation of all their work. Stability,
however, was the word they used, not
immutability. Classification was their
special work, and the whole principle of
classification, as Mr. Spencer truly says,
rests on the idea, and on the fact, that
all living creatures can be arranged in
groups by endless cycles of definite

affinity and of definite divergence. Linnæus applied this principle to the living world as it exists now, and his famous Binomial system, which survives to the present day, assumes, as a fact, that in that world genera and species are practically stable. Cuvier, on the other hand, was largely concerned with the extinct forms of life, and his classification of them, and his identification of their relations with living forms, would have been impossible if the peculiarities of the structure in all living things had not maintained through unknown ages the same persistent character. He therefore declared, with truth, that the very possibility of establishing a science of natural history absolutely depends on the stability of species.

If, then, we give up the idea that species have been permanently immutable, we must beware of rushing off to antithetical conclusions which are at variance with at least all contemporary facts in the living world, and which, as regards the past, rest mainly on our impossibilities of conception in a matter on which we are profoundly ignorant. Species, if not absolutely immutable, have now undoubtedly, and always have had, a very high degree of stability and endurance.   If mutations have occurred, it must have been under some conditions, and under some law, of which we have no example and can form no conception. It is at this point that the theory of organic evolution, when understood in what may be called the party sense, breaks down as an easy explanation of the facts.   It may be true that the idea of separate creations continually repeated is an idea which represents an escape from thought, rather than an exercise of reasonable speculation on the processes through which development has been conducted. But exactly the same may be said of the idea of species being so unstable that they were constantly passing into each other by nothing but fortuitous and infinitesimal variations.

This, indeed, may be an easier and lazier conception than any other. But it is easier only because it takes no notice of insuperable difficulties and disagreements with the facts. Species have been quite as stable throughout all the geological ages as they are at present. Linnæus's Binomial system of classification is as applicable to, and fits as well into, the Trilobites of the Palæozoic rocks —the Brachyopods and the Cephalopods of the Secondary ages—the Mammalia of the Tertiary epoch, as it fits into all the species now alive or only recently extinct. Each species has its own dis-

tinctive characters, down to the minutest
ornamentation on a scale or on an
osseous scute, or to the peculiar varieties
of pattern on the convolutions of an
Ammonite. These species continue till
they die, and then they are often suddenly
replaced by new forms and new patterns,
all as definite and as persistent as before.
How this takes place no man as yet can
tell.

I recollect one striking illustration.
Some thirty-five years ago I visited the
distinguished French geologist Barrande,
who devoted himself for years to the life-
history of the Trilobites in the Silurian
rocks of Bohemia. He had a magnificent collection of those curious crustaceans in his house in Prague. Nothing
was more remarkable than the stability
of the forms which he identified. This
stability extended to the immature or
larval forms of each species. He had

specimens in every stage of growth. He was good enough to drive with me to the beds of rock which contained them. They were the rocks forming in low but steep hills—the containing walls of the Valley of the Moldau. They consisted of a highly fissile slaty rock, the planes of which were often charged with the fossils. They seemed to me to be singularly regular and unbroken by clefts or chasms; yet in the middle of these regular and consecutive beds there were members of the series which suddenly displayed new species. Barrande was puzzled by the phenomenon. Where could these new species come from? It never occurred to him that possibly they might be born suddenly on the spot. So, to meet the difficulty, he invented the theory of "colonies" — emigrants from some other centre which had migrated and settled there. Of course,

this is no solution, but only a banish-
ment of the difficulty to some other
place.   The more common bolt-hole for
escaping from this difficulty is to plead
the "imperfection of the record."   But
this does not really avail us much.   As
regards terrestrial forms of life, indeed,
it is true that the record is very imperfect, because the conditions are rare and
partial under which land animals can be
preserved in aqueous deposits.   Consequently, as regards them, we never get
a complete series.   But there are many
great rock-formations of marine origin,
which were continuous deposits for ages,
at least long enough to embrace the
first appearance of many new species.
Yet these new species never seem to
be mere haphazard variations from pre-
existing forms.   They never have the
least appearance of the lawless mixtures
of hybridism.   On the contrary, the

new forms are always as sharply defined
as the old, differing from them by characters which are as well marked and as
constant as all their predecessors in the
wonderful processions of organic life.   It
helps us very little to remember that in
the existing world some varieties do
occur in certain species—varieties which
are sometimes sufficiently well marked
to raise the question among classifiers
whether they are, or are not, sufficiently
constant to deserve the name of separate
species.   But this does not help us
much, because such varieties are very
limited in extent, and are almost always
confined to such superficial features as
the colour of hair or of feathers.   They
never, so far as I know, affect organic
structure, and no accumulation of them
would account for the very different
kinds of variation which are conspicuous
in the successions of organic life.

These, however, are not the only
difficulties which beset any intelligent
acceptance of the theory of purely mechanical and mindless evolution through
changes infinitesimal and fortuitous.
There is another difficulty much more
fundamental. That theory, in all its
forms, involves always one assumption,
which, so far as I have observed, is
never expressly stated. It is the assumption that organic life never could have
been introduced, or multiplied, except by
the processes of parental reproduction or
of ordinary generation, such as we see
them now. Yet—if we only think of it
—this is an assumption which not only
may be wrong, but which cannot possibly be true. We know as certainly
as we know anything in the physical
sciences, that organic life must have had
a definite beginning, in time, upon this
globe of ours. If so, then of course that

beginning cannot possibly have been by way of common parentage or ordinary generation. Some other process must have been employed, however little we are able to conceive what that process was. All our desperate attempts, therefore, to get rid of the idea of creation, as distinguished from mere procreation, are self-condemned as futile. The facts of Nature, and the necessities of thought, compel us to entertain the conception of an absolute beginning of organic life, when as yet there were no parent forms to breed and multiply.

Darwin, as is well known, recognised this ultimate necessity. He clothed the conception of it in words derived from the old and time-honoured language of Genesis. He spoke of the Creator first breathing the breath of life into a few, perhaps only into one single organic form. His followers generally seem to regard this as a weak concession on the part of their great master. Darwin himself, in a letter to Sir J. Hooker, was weak enough to express his own regret. And yet he went on publishing edition after edition without changing his words or withdrawing them, or offering any explanation, or suggesting any alternative conception. And why? Because he had none to suggest. His followers are generally silent on the significance of this passage in their master's intellectual experience. His instinct that life must have had a beginning, as subversive of the fundamental assumption of his theory, they pass over in silence. They never dwell on it. They never realise that without it, or without some substitute for it, the whole structure of what they call organic evolution is without a basis—that it represents a chain hanging in mid air, having no point of attachment in the heavens or on earth. It is as certain as anything in human thought that, when organic life was first introduced into the world, something was done—some process was employed—differing from that by which those forms do now simply reproduce and repeat themselves.

But the moment this concession has been fully, frankly, and intelligently made, another concession necessarily follows, namely this, that we cannot safely conclude that the first, and more strictly creative, process has never been repeated. Yet this is the assumption tacitly involved in all the current materialistic theories of evolution. They all absolutely depend upon it, although it is seldom if ever avowed. It is an assumption, nevertheless, in favour of which there is assuredly no antecedent probability. On the contrary, the true

presumption is that, as solitary excep-
tions are really unknown in Nature, the
same processes may very well have been
often repeated from time to time. Or
perhaps even it may be true that such
processes are involved in, and form an
essential part of, the infinite mysteries of
what we call, and think of so carelessly,
as ordinary generation. This is an idea
which opens very wide indeed our intellectual eyes, and gives them much to do
in watching and interpreting the fathom-
less wonder of familiar things.

Let us, however, provisionally at least,
accept the belief that organic life was first
called into existence in the form of some
three, or four, or five germs—each being
the progenitor of one of the great leading types of the animal creation in
respect to peculiarities of structure—one
for the Vertebrata, one for the Mollusca,
one for the Crustacea, one for the

Radiata, and one for the Insecta. Let
us assume, farther, on the same footing,
that from each of these germs all the
modifications belonging to each class
have been developed by what we call
the processes of ordinary generation.
Then it follows that, as all these modifications have undoubtedly taken definite
directions from invisible beginnings to
the latest results and complexities of
structure, the original germs must have
been so constituted as to contain these
complexities, potentially, within themselves.   This conclusion is not in the
least affected by any influence we may
attribute to external surroundings.   The
Darwinian school in all its branches invariably dwell on external conditions as
physical causes.   But it is obvious that
these can never act upon an organic
mechanism except through, and by means
of, a responsive power in that mechanism

itself to follow the direction given to it, whether from what we call inside or outside things.

This is no transcendental imagination, as some might think it. It is a conclusion securely founded on the most certain facts of embryology. It is the great peculiarity of organic development or growth that it always follows a determinate course to an equally determinate end. Each separate organ begins to appear before it can be actually used. It is always built up gradually for the discharge of functions which are yet lying in the future. In all organic growths the future dominates the present. All that goes on at any given time in such growths has exclusive reference to something else that has yet to be done, in some other time which is yet to come. On this cardinal fact, or law, in biology there ought to be no dispute with Mr. Spencer. Numberless writers before him have indeed implied it in their descriptions of embryological phenomena, and of the later growth of adapted organs. But, so far as I know, no writer before Mr. Spencer has perceived so clearly its universal truth, or has raised it to the rank of a fundamental principle of philosophy. This he has done in his *Principles of Biology*, pointing out that it constitutes the main difference between the organic and the inorganic world. Crystals grow, but when they have been formed there is an end of the operation. They have no future. But the growth of a living organ is always premonitory of, and preparative for, the future discharge of some functional activity. As Mr. Spencer expresses it, " changes in inorganic things have no apparent relations to future external events which are sure, or likely, to take place. In vital changes, however, such relations are manifest."[1] This is an excellent generalisation. It only needs that the word "relations" be translated from the abstract into the concrete. The kind of relation which is "manifest" is the relation of a previous preparation for an intended use. Unfortunately, Mr. Spencer is perpetually escaping or departing from the consequences of his own "manifest relations." In a subsequent passage of the same work[2] he says, "Everywhere structures in great measure determine functions." This is exactly the reverse of the manifest truth — that the future functions determine the antecedent growth of structure. This escape from his own doctrine on the fundamental distinction between the organic and the inorganic

[1] Spencer's *Principles of Biology*, vol. i. ch. v. p. 73.
[2] *Ibid.* vol. ii. ch. i. p. 4.

world is an escape entirely governed by his avowed aim to avoid language having teleological implications. But surely it is bad philosophy to avoid any fitting words because of implications which are manifestly true, and are an essential part of their descriptive power.

If, therefore, we are to accept the hypothesis that all vertebrate animals, whether living or extinct, have been the offspring, by ordinary generation, of one single germ, originally created, then that original germ must have contained within itself certain innate properties of development along definite lines of growth, the issues of which have been forearranged and predetermined from the first. I have elsewhere[1] shown how this conception permeates, involuntarily, all the language of descriptive science when specialists take it in hand

[1] *Philosophy of Belief*, ch. iii.

to express and explain the facts of Biology to others. Huxley habitually uses the word "plan" as applicable to the mechanism of all organic frames.

This is a theory of creation—by whatever other name men may choose to deceive themselves by calling it. It is a theory of development too, of course, but of the development of a purpose. It is a theory of evolution also—but of evolution in its relation to an involution first. Nothing can come out that has not first been put in. It is not less a theory of creation which, whether true or not, gets rid absolutely of the elements of chance so valued by Darwin's more fanatical followers, and of the mere mechanical necessity which seems to be favoured by Mr. Spencer.

It must be obvious, however, that the burden of this conception would be greatly lightened if we give up the un-

justifiable, and indeed irrational, assumption that what must confessedly have
happened once can never possibly have
happened again, namely, the introduction of new germs with their own special
potentialities of development.   There
are natural divisions in the animal kingdom which seem to suggest the idea of
a fresh start on new lines of evolution.
The Mammalia may well have been
thus begun as a great advance on the
hideous Reptiles, which once dominated
the world both by land and sea.   Fishes
may well have had another separate
ancestral germ — and so with all the
lower orders of creation, some of which
are very deeply divided from each other.
I know of no natural or rational limita-
tion on the possibilities of this suggestion.   On the contrary, the general law
of the continuity of Nature is favourable
to repetition of any and every precedent

which has once been set in the processes
of creation. There is an antecedent probability that anything done once has been
done again and again—that, in fact, it is
part of a system, and in fulfilment of a law.

The conceivableness of this process
would be indefinitely increased if we
invoke the help of another principle,
and of another analogy in the actual
phenomena of organic life—and that is
the great rapidity with which organic
germs can sometimes evolve their in-
volutions—and develop their predestined
and prearranged adaptitudes. The Darwinian idea has persistently been that
the steps of development have been
always infinitesimally small, and that
only by the accumulation of these,
during immeasurable ages, could new
forms have been established. It has
long occurred to me that this assumption is against the analogies of Nature,

seeing that in all cases of ordinary generation, and conspicuously in a thousand cases of metamorphoses among the lower creatures, the full development of germs takes a very short time indeed. In the case of some birds, a fortnight or three weeks at the outside is sometimes enough of time wherein to develop, from an egg, a complete fowl with legs, and wings, and instincts, all ready-made to lead an adult and independent life. In frogs and toads the time of hatching varies from three days to three weeks. In some insects a few hours is enough to produce a creature very highly organised, with many special adaptations. In other numberless cases, a living creature, already leading a separate life, is put to sleep within an external case or shell, and, in that state of sleep, is radically transformed in all its organs, and comes out in a few days an entirely new animal form, with new powers, fitted for new spheres of activity and of enjoyment. All these incomprehensible facts —in which nothing but the blinding effects of familiarity conceals from us the really creative processes involved — demonstrate the absurdity of supposing that new species could not be evolved from germs except by steps infinitesimally slow, and accumulated through unnumbered ages.

This powerful argument, securely founded on the most notorious facts of the living world, has for many years entirely relieved my mind from the supposed difficulty of reconciling all that is essential in the idea of creation with the pretended competing idea of evolution or development. I have not, however, hitherto used it publicly, not having had a fitting opportunity of so doing. But I do not recollect having seen it used by others. It is, therefore, with

no small surprise that, in Mr. Spencer's
article, I find it taken up and used for
a wholly different contention. His adoption of it is a good example of the uses
of controversy.   Thirty-two years ago
he would not have used it.   We have
good evidence of this in a vigorous
letter published in the Appendix to vol. i.
of his *Principles of Biology*, 1864.   In
that letter he makes "enormous time"
an essential condition of even the very
lowest steps in organic evolution.   And
for a good reason, which, with his usual
candour, he frankly explains.   The
sudden or very rapid evolution of even
the lowest organic forms, from some
primordial germs, he sees plainly, would
be a very dangerous admission. "If,"
he says, "there can suddenly be imposed
on simple protoplasm the organisation
which constitutes it a *Paramœcium*, I
see no reason why animals of greater

complexity, or indeed of any complexity,
may not be constituted after the same
manner." Neither do I. Therefore, to
escape from an idea so perilous to his
philosophy, he asserts his conviction
that "to reach by this process (organic
evolution) the comparatively well-special-
ised forms of ordinary *Infusoria* must
have taken an enormous period of time."[1]
To find, therefore, Mr. Herbert Spencer
now insisting on the actual rapidity, and
the still greater conceivable rapidity,
of evolution in organisms, is a very
instructive change of front. It is for
the sake of argument that he takes
up this new attitude on an all-important point. Lord Salisbury in his Address had dwelt on the immensities of
time which, on the Darwinian theory,
must have been needed to develop
"a jelly-fish into a man"; and he had

[1] P. 481.

confronted this demand on time with the calculations of physicists, which limit the number of years since the globe must have been too hot for organic life. I have never myself dwelt on this objection to Darwinism, because I never felt absolute confidence in the calculations of decreasing heat which vary from tens of millions to hundreds of millions of years. Recently, however, Lord Kelvin has placed it on strong grounds of calculable certainty that the demands of many geologists on time have been extravagant and impossible. Still, when we get into such high numbers as even twenty millions of years, and such enormous margins for possible error, I always feel that we are handling weapons which have no certain edge. But Mr. Spencer now avails himself of the safer alternative when he escapes from the difficulty by throwing overboard altogether the

doctrine that changes in animal structure
can only have been very minute and very
slow.   He, therefore, takes up the same
idea that has often occurred to me—that
all the phenomena, even of ordinary
generation, point to the possibility of
great transmutations having been accomplished in very short periods of time.
It seems he had foreshadowed this line
of argument in 1852, before Darwin's
book was published.   But he now works
it out in more detail, and revels in the
calculations which prove what great
things are now being very summarily
done by ordinary generation in develop-
ing the most complex organic forms
from a simple cell.   The nine months
which are enough to develop the human
ovum into the very complex structure of
a new-born infant are divisible, he calcu-
lates, into 403,200 minutes.   If only one
hundred millions of years were allowed

since the globe was cool enough to
allow of life, then, he argues, no less
than 250 years would be available out
of each minute of man's development—
for those analogous changes which have
raised some Protozoon into Man. Mr.
Spencer makes no mention of the conspicuous wonders effected in insect and
crustacean metamorphoses during periods
relatively much shorter. He makes no
allusion to the fact that specialists often
speak of embryonic stages, common in
some genera, being "hurried over" in
the case of others, so that the final
stages are more quickly reached. An
idea so suggestive of a directing and
creative energy thus visibly subordinat-
ing the machinery of generation to
special ends, is an idea which goes far
beyond Mr. Spencer's new argument
deprecating the over-importance hitherto
attached by thoughtless evolutionists to

K

countless ages of infinitesimal change. He may well say that if this be true, no reason can be seen why animals of any degree of complexity may not be developed as quickly and after the same manner.   Neither, of course, does Mr. Spencer push his argument to the obvious conclusion which is adverse to his philosophy—the conclusion, namely, that if the first creation of germs has ever been repeated, still more if it may have been frequently repeated, then the whole processes of a creative development may have been indefinitely hastened, and the element of time becomes of quite subordinate importance.

CHAPTER III

CLUES AND SUGGESTIONS

MR. HERBERT SPENCER'S rejection of any
necessity for the "enormous" time which
evolutionists have hitherto demanded,
and to which Lord Salisbury only
alluded as a well-known characteristic
of their theories, marks a new stage in
the whole controversy. Nobody had
made the demand more emphatically
than Mr. Spencer himself only a few
years ago. His confession now, and
his even elaborate defence of the idea
that the work of evolution may be a
work of great rapidity, goes some way
to bridge the space which divides the

conception of creation, and the conception of evolution as merely one of the
creative methods.   But Mr. Spencer
must make further concessions.   It is not
the element of time, however long, nor
is it the mere idea of a process, however
purely physical, which we object to—
we who have never been able to accept
any of the recent theories of evolution
as giving a true or adequate explanation
of the facts of organic life.   The two
elements in all those theories which we
reject as essentially erroneous, are the
elements of mere fortuity on the one
hand, and of mere mechanical necessity
on the other.   If the processes of
ordinary generation have never been
reinvigorated by a repetition of that
other process—whatever it may have
been—in which ordinary generation was
first started on its wonderful and
mysterious course, then all the more

certainly must the whole of that course have been foreseen and prearranged. It has certainly not been a haphazard course. It has been a magnificent and orderly procession. It has been a course of continually fresh adaptations to new spheres of functional activity. We deceive ourselves when we think or talk, as the Darwinian school perpetually does, of organs being made or fitted *by* use. The idea is, strictly speaking, nonsense. They must have been made *for* use, not *by* use, because they have always existed in embryo before the use was possible, and, generally, there are many stages of growth before they can be put to use. It is, therefore, a fact—not a theory—that during all these stages the lines of development were strictly governed by the end to be attained—that is to say, by the purpose to be fulfilled.

This, indeed, is evolution; but it is the evolution of mind and will; of purpose and intention. We are not to be scared by the application to this indisputable logic of that most meaningless of all words—the supernatural. For myself I can only say that I do not believe in the supernatural—that is to say, I do not believe in anything outside of what men call Nature, which is not also inside of it, and manifest throughout its whole domain. I cannot accept, or even respect, the opinion of men who, in describing the facts of Nature, and especially the growing adaptations of organic structures, use perpetually the language of intention as essential to the understanding of them, and then repudiate the implications of that language when they talk what they call science or philosophy. When evolutionists do defend their

inconsistencies in this matter, they use
arguments which we cannot accept as
resting on any solid basis. Thus Mr.
Spencer argues in his article that if the
Creator had willed to form all those
creatures, He surely would have led
them along lines of direct growth from
the germ to the finished form, and would
not have led them through so many
stages of metamorphoses.[1] We have
no antecedent knowledge of the Creator
which can possibly entitle us to form
any such presumption as to His methods
of operation. This is one answer. But
there is another. The method which is
supposed by Mr. Spencer to be inconsistent with the operations of a mind
and will is the same method which is
our own, and which is universally prevalent in the Universe. Everything is
done by the use of means; everything

[1] P. 745.

is accomplished by steps, generally visible, but often also concealed from our view. There is, therefore, either no mind guiding the order of that universe, or else this method is compatible with intellectual direction. We must take Nature as we find it. We have nothing to do with what Mr. Spencer calls "Special Creation." Special evolution will do very well for our contention. That contention is that in organic structures purposive adaptations have had the controlling power. This is not an argument; it is a fact. In Biology our perception of the relation between organic structures and the purposes they are made to serve—which are the functions they are constructed to discharge—is a perception as clear, distinct, and certain as our perception of their relations to each other, or to time, or to form, or to space, or to any other of the categories of our knowledge.

Mr. Spencer is under a complete delusion if he supposes that the four or five great heads of evidence, which he specifies as all telling the same tale of evolution, could not be equally applicable to the facts if all the steps of evolution were visibly and admittedly under the ordering and guidance of a will. For example, the argument founded on the possibilities of Classification applies to the evolution of human machines as well as to the organic mechanisms of Nature. A row of models of the steam-engine, from "Papin's Digester" to the wonderful machines which now drive express trains at sixty or seventy miles an hour, would show a consecutive series of developments in every way comparable— except in length and complexity—with the series of the Mammalian skeleton.

Yet nobody would be tempted to guess on this account, except in a metaphorical sense, that steam-engines have all been begotten by each other.   The metaphor from organic births, however, is so apposite and perfect in its analogy that it is often actually used, and the begetting of ideas, or of the application of ideas to mechanical or chemical work, is a recognised branch of the history of mechanics.

The truth is that the argument derived from the principle on which all natural classifications rest, is a very dangerous argument for Darwinians. It cuts two ways, and one of the ways is very undermining to the assumption that there has been some continual flux of specific characters.   It is true that in all living structures common features, so numerous, do indicate some common cause and source.   But it is not less true that specific differences, so constant and so definite through enormous periods of time, are incompatible with perpetual instability.   Darwin himself spoke of "fixity" as an essential characteristic of true species.   He admitted that this fixity is never attained by the human breeder; and he even admitted that it could only be obtained by "selection with a definite object."[1]   This is a most remarkable declaration.   Just as we have seen Mr. Spencer, under the inducements of controversy, throwing overboard his old demand for enormous periods of time, so now we find Darwin throwing overboard the idea of variations being either constant, or indiscriminate, or accidental, and even insisting that "fixity" in organic forms is an aim in Nature, and can only be secured through

[1] Quoted by Professor Poulton, *Charles Darwin*, etc., p. 201.

an agency having a definite object, and pursuing that object with a persistency impossible to man as a mere breeder of temporary varieties. This is an argument which gives a very high rank to species in the history of life. It is because of it that Cuvier declared that no science of Natural History is possible if species be not stable. If, then, it be true that one species has always given birth to others, it must have been by a process of which, as yet, we know nothing.

And then it must be remembered that there are some fundamental features in all living organisms—involving corresponding likenesses—which can have no other than a mental explanation. One great principle governs the whole of them, namely this, that in order to take advantage of special laws, physical, mechanical, chemical, and vital, certain

corresponding conditions must be sub-
mitted to, and certain apparatuses must
be devised, and provided, for the meeting of these necessities. But the bond
—the nexus—between the existence of
a need and the actual meeting of that
need, in the supply of an apparatus, can
be nothing but a perceiving mind and
will. I quite agree with Mr. Spencer
that most men when they talk of separate
or special Creation do not realise, or
"visualise," what they mean by it. But
exactly the same criticism applies to the
language of those who are perpetually
explaining organic structures as developments governed by the absolute necessities of external adaptations. They do
not really see the necessary implications
of their own language. If the organism
is to live at all, they frequently tell us,
such and such developments must arise.
Quite so—but who is it, or what is it,

that determines that the organism shall live, and shall not rather die? The needed development will not appear of its own accord. The needed perception of its necessity must exist somewhere; and the needed power of meeting that necessity must exist somewhere also. Moreover, the two must act in concert. Those, therefore, who talk about that combined perception and power existing in Nature are using words with no meaning, unless by Nature they mean a conceiving and a perceiving agency. It is on this principle alone that we can explain very clearly why certain lines of structure and certain special apparatuses are common to all living things. The assimilation of food,—the support of weight,—some fulcrum for the attachment of muscle,—some circulatory fluid, —some vessels for the circulating fluids to find a channel,—some apparatus for the supply of oxygen, and for its absorption, —some nervous system for the generation of the highest energies of life,— some optical arrangement for the purposes of sight—all of these, and many more, involve, of necessity, likenesses and correspondences between all living things in the animal kingdom. These correspondences hang together by a purely mental and rational chain of common necessities which have been seen and have been accordingly provided for. These mental relations between needs and their supply are entirely independent of the methods employed, and, as a fact, the methods employed do very considerably vary. The argument would be exactly the same if the methods of supply were much more various than they actually are. If the one method employed has never been anything but ordinary generation,—with the single

exception of the first, or the few first, of
the whole series,—then it would follow
that the amount and the definiteness of
the prevision involved in the first germs
must have been all the more wonderful,
and the more completely answering to
all that can be intelligible as creation.

There is surely something suspicious
—improbable—at variance with all the
analogies of Nature—in the doctrine
which the mechanical evolutionists would
force upon us — that the life - giving
energy, by whatever name we may call
it, which started organic life upon its
way—in the form of some four or five
primordial germs — has been doing
nothing ever since. No doubt it mag-
nifies the richness and fertility of the
original operation—seeing as we do the
almost infinite varieties which it included
in its predetermined lines of change.
But if this has been the course of creation,

we are driven to another conception without which the theory would not at all correspond to the facts of life. If ordinary generation has been the sole agent in producing all but the few original germs, then ordinary generation must have been sometimes made to do some very extraordinary things. Mr. Spencer very fairly admits that man has never yet seen a new species born by ordinary generation. This may be theoretically accounted for by the shortness of man's life as yet upon the globe. But, unfortunately for the theory, the long ages of Palæontology give no clue to the immediate parentage of any new species. There are, indeed, intermediate forms, and these are called links. But somehow the links never seem to touch. The new forms always appear suddenly —from no known source—and generally, if of a new type, exhibiting that type in

L

great strength as to numbers, and in
great perfection as regards organisation.
The usual way of evading this great
difficulty in the facts of Geology is to
plead what is called the imperfection of
the Record. But this plea will not avail
us here. There are some tracts of time
respecting which our records are almost
as complete as we could desire. In the
Jurassic rocks we have a continuous and
undisturbed series of long and tranquil
deposits—containing a complete record
of all the new forms of life which were
introduced during these ages of oceanic
life. And those ages were, as a fact,
long enough to see not only a thick
(1300 feet) mass of deposit, but the first
appearance of hundreds of new species.
These are all as definite and distinct from
each other as existing species. No less
than 1850 new species have been counted
—all of them suddenly born—all of them

lasting only for a time, and all of them
in their turn superseded by still newer
forms. There is no sign of mixture, or
of confusion or of infinitesimal or of in-
determinate variations. These " Medals
of Creation " are all, each of them, struck
by a new die which never failed to im-
press itself on the plastic materials of this
truly creative work. There is nothing
more instructive than to place a series
of these new species, such as the Ammon-
ites, on a table side by side. The perfect
regularity and beauty of each new
pattern of shell, and the fixity of it so
long as it existed at all, are features as
striking as they are obvious.

There is one suggestion which has
been made in order to meet these strange
phenomena, which has always seemed
to me to be more plausible than any
other, and to come much nearer than
any other to the historic facts. It was

the suggestion of a very eminent and most ingenious man—Babbage, the inventor of the Calculating machine. His mind was full of the resources of mechanical invention. He conceived the idea that as such a machine as his own could be made to evolve its results according to a certain numerical law during a given time, and then suddenly, for another time, to follow a different law with the same accuracy and perfection of results, so it is conceivable that species might be really as constant and invariable as we actually find them to be, for some long periods of time—embracing perhaps centuries or even millenniums—and then suddenly, all at once, evolve a new form which should be equally constant, for another definite time to follow.

This notion would account for many facts, and it is, of course, consistent with the assumption that what we call ordinary generation has—since in the first creations it was originally started on its way—been the only and the invariable instrumentality employed in the development of species. And not only would this idea square with the apparently sudden appearance of new species, repeated over and over again throughout the geological ages, but, more important still, it would harmonise with those intellectual instincts and conceptions of our mental nature to which the idea of chance is abhorrent, and which demand for an orderly progression in events some regulating cause as continuous and as intelligible as itself.

Mr. Spencer refers, as others now continually do, to the recent discoveries in America which have revealed a remarkably continuous series of specific forms leading up to that highly specialised animal the Horse. That series of forms, although then less continuous, was noticed long before the days of Darwin. It attracted the attention of Cuvier, and I heard Owen lecture upon it as indicative of the origin of the Horse two years before the *Origin of Species* had been published. The later more near approach to completion in that series of American fossils is said by Mr. Spencer to have finally convinced Professor Huxley of conclusions on which he had before maintained a certain reserve. They are, indeed, most significant, but I am not sure that their significance has been well interpreted. They do indeed seem to indicate the development of a plan of animal structure worked out, somehow, through the processes of ordinary generation. But they do not indicate any fortuity, or any confusion, or any blind haphazard variations in all possible

directions.   Neither do they indicate
steps of infinitesimal minuteness.   On
the contrary, they indicate a steady progress in one determinate line of development, a progress so rapid that sometimes
the new species seem to have been
actually living as contemporaries with
the older species ; and alongside of the
anterior forms which were, as it were,
going out of fashion, and are now assumed
to have been their own progenitors.
The number, too, of the forms through
which the line of modifications can be
traced during a geological period of
apparently no long duration, indicates at
that time an activity in the production of
new specific characters which is highly
suggestive of comparatively rapid changes
in the processes and in the products of
ordinary generation.   Sedimentary beds
not exceeding 180 feet in total thickness,
and thus indicative of no very long time

in the geological scale, are now found to contain several of the divergent forms which lead up to the fully developed Horse.[1]   It is as if the creative energy, which on every theory must have begun the series in the creation of the original germs, had been then calling out their included potentialities into manifestations unusually rapid.   These manifestations were all pointing steadily in one direction, namely, the establishment—on a continent ceasing to be marshy—of a species of quadruped, organised for a singular combination of strength, and fleetness, and endurance in the machinery of locomotion upon drier land.

This example of the correlations of growth effected in all probability through

[1] I have taken these facts from a very remarkable paper in the *Proceedings of the American Philosophical Society* for August 1896, "On the Osteology of the White River Horses," by Marcus S. Farr, pp. 147-175.

the machinery of ordinary generation, but under a definite guidance along certain lines to an extraordinary but determinate result, is all the more striking because it does not stand alone. All the great domesticable Mammalia which serve such important purposes in the life of Man, and without which that life would have been far less favourably conditioned than it is, were all the contemporaneous product of that very recent, but most pregnant, Pliocene age in which the Horse was, at some appointed time, evolved out of ancestral forms, which would have been as useless to Man as the survivors of them now are, such as the Rhinoceros or the Tapir.

Among the conceptions to which the Darwinian theory of development has most frequently resorted, has been the conception that the development of all individual things from germs is an

epitome and an analogue of the kindred,
but far slower and longer, processes
which have given birth to species in the
course of ages. It is the best of all their
conceptions—that which most facilitates
the imagination in picturing a possible
method of creation—because it rests on
at least a plausible analogy of Nature.
But, unfortunately, the mechanical school
of evolutionists do not seem to under-
stand one of the most certain characteristics of the processes of ordinary generation. If the germs first created had all
the essential qualities of the procreated
germs, then chance, or miscellaneous
and unguided growths, can have had no
place in the development of species.
Nothing can be more certain that every
procreated germ runs its own peculiar
course to its own peculiar goal, with a
regularity that implies a directing force.
Mr. Spencer himself reminds us that all

procreated germs are so like each other
in the earliest stages, that neither the
microscopist, nor the chemist, could tell
whether any germ is to develop into
any of the lowest animals or into a man.
Yet the line of growth, in each, is pre-
determined, and the adult form is as
certain and as definite as if the completed
animal had been a separate creation from
the inorganic elements of Nature. If,
therefore, the mechanical evolutionists
appeal to the processes of ordinary
generation, they must take all the consequences of that appeal. They must
not reject or gloss over a feature of it
which is most fundamental and conspicuous, namely, the internal directing agency
or force, which always pursues a definite
line of growth, so that all the demands
of the completed structure must have
been present from the beginning, and
must have been always ready to appear

in strength when the set time had come, and very probably to appear in embryo even sooner.

It has always appeared to me that this is a conception of such strength, and even of such certainty, that it casts a new and a very clear light on one of the most curious and puzzling groups of fact which the science of Biology reveals— I allude to the frequent occurrence in animal structures of what are called rudimentary organs—that is to say, the occurrence of bits of organic mechanism which are never to be used in that particular creature, but which, in other creatures widely different, grow up into functional activity, and may even be the most essential organs of its life. A great number of instances have been cited by comparative anatomists—some of them, perhaps, more fanciful than real—as, for example, when the five or six vertebræ

which constitute a real, though an invisible, tail in Man, are quoted as a case
of a rudimentary organ.   The truth is
that this very short tail in men is far
more clearly functional than many very
long tails in other animals.   It is
absolutely needed for the support of the
whole frame when it is subjected to the
strain of its own weight for long periods
of time in the sitting posture, a posture
which is peculiar to Man and, in a less
degree, to Monkeys.   It is not clear that
there is any functional use in the long
tails of dogs, of cats, and of many other
animals.   They are, indeed, very ex-
pressive of the emotions, and this, no
doubt, is of itself a use.   Perhaps more
really belonging to the category of rudi-
mental organs may be the traces which
are said to exist in the human head of
the special muscles which move the ears
in lower animals.   If such exist, although

a certain very limited power of move-
ment of the scalp is observable in a few
individuals, such muscles seem to be
divorced in man from their appropriate
use.

But it is needless to dwell on cases
which can only be verified by specialists
in anatomy, when we have in Nature
conspicuous cases which, when seen,
confront us with perpetual but baffled
curiosity and astonishment.   The most
extreme case is the best for illustration,
and is naturally the most often quoted.
It is the case of the Whale.   This hugest
of all the living vertebrata is so exclusively adapted to life in the ocean that if
by accident it is stranded on the shore
it is speedily suffocated by the crushing
of all its internal organs under its own
enormous weight.   Yet this creature, so
utterly destitute of any osseous structure
capable even for a moment of sustaining

that weight, does, nevertheless, exhibit in its skeleton all the bones which constitute the fore limbs of quadrupeds, and has even a bony rudiment which represents the elaborate structure which, in them, constitutes the pelvis. This is the solid fulcrum upon which, in them, the posterior pair of limbs are hinged, and on which, in the case of Man, the power of progression on land is absolutely dependent. The Whale, too—at least that species of whale called the Right Whale, which is the species we know best, from its great commercial value—presents in its life-history another example of rudimentary organs. The new-born whale is provided with teeth, which are utterly without functional use either in the young or in the adult, and are soon absorbed and lost as the young advance to maturity.

There is no doubt that the class of

facts to which these belong are guide-
posts in the science of Biology.   They
must have an historical origin, and a
meaning, which is not yet thoroughly
understood.   Let us look at some considerations which seem to throw an
important light upon them.

In the first place, it is evident that
organic structures, or bits of organic
structure, which have no apparent use
at all to some individual creatures possessing them, are closely connected with
that other case which is much more
common—the case, namely, of the same
organic structures existing in different
animals, but which are in them put to
entirely different uses.   Owen says that
even the cetacean pelvis is used, in the
meantime, for the attachment of some
muscles connected with the generative
organs.   The five digits of a man's
hand, again, are identical in number and

position with the five slender bones of a Bat's wing. In that animal they are used as the supporting framework of a flying membrane, and are wholly useless for any purposes of prehension. The digit which we call our thumb, and which in Man has such essential uses that the hand would hardly be a hand without it, is in the Bat not altogether abolished, but is dwarfed and converted into a mere hook by which the creature catches hold of the surfaces to which, when at rest, it clings. The whole vertebrate creation is full of such examples. Rudimentary organs, therefore, are nothing but a natural and harmonious part of a general principle which is applied in different degrees throughout the animal world. The explanation is, in one sense, very simple. It is that the vertebrate skeleton, with all its related tissues, has been—what Huxley

always called it—a Plan, laid down from
its beginning, in its originating germs,
with a prevision of all its complexities
of adaptability to immense varieties of
use.   There must have been a provision
for these uses in certain elements and
rudiments of structure, and in certain
inherent tendencies of growth, which
were to commence, from time to time,
the new and specially adapted structures.
This is the indisputable fact in every
case of ordinary generation, and if that
process has been the only method employed since the first few germs were
otherwise created, then both the cause
and the reason of rudimentary organs
in many creatures become intelligible
enough.

There is nothing in this explanation
which can be rationally objected to by
evolutionists.   Indeed, if Darwin's particular theory of development be at all

true, it becomes an absolute necessity of thought that there must have been, in the history of organic life, a whole series of special organs appearing from time to time as rudiments, and then, after a period of functional activity, disappearing again as vestiges. The course of organic life has certainly been, on the whole, one of progress from lower to higher organisations, and if it be true that all these changes have come about with infinitesimal slowness—or even if they have been occasionally rapid—there must have been always as many structures in course of preparation for future use, as there were other structures in course of extinction because they were ceasing to be of any use whatever.

It is curious to observe that Darwinians, generally, never seem to perceive this necessity at all. When they see a rudimentary organ in any animal frame they always insist that it must be the vestige of an organ which was once in full activity in some actual progenitor. They never allow that it may possibly represent a contemplated future. According to them it must, and can, only represent an accomplished and concluded past. Why is this? Of course it involves a complete abandonment of the attempt to give any account of the origin of any organic structure. It implicitly assumes that they were created suddenly, and in a state so perfect as to be capable of functional activity from the moment of their first appearance. If not, then there is no puzzle in rudimentary organs. They are the normal and necessary results of gradual evolution by gradual variations.

The assumption, therefore, that such organs must always be the remnants of structures formerly complete, is so

entirely at variance with the whole
theory of the mechanical evolutionists
that there must be some explanation
of their running their heads against it.
The explanation is very simple. It
is one of the infirmities of the human
mind that, when it is thoroughly
possessed by one idea, it not only sees
everything in the light of that idea, but
can see nothing that does not lend
itself to support the dominant conception. There is nothing that a mind
in this condition dislikes so much as an
incongruous fact. Its instincts, too, are
amazingly acute in scenting, even from
afar, the tainted atmosphere of phenomena which have dangerous implications. This is the secret of the aversion
felt by the Darwinian School to the
immense variety of biological facts which
point to the steady growth of organs for
a predestined use, and consequently to

their inevitable first appearance in rudimentary conditions in which as yet they
can have no actual functional activity.
For this is an idea profoundly at variance
with materialistic and purely mechanical
explanations. It is easy by such explanations—at least superficially it seems
to be easy—to explain the atrophy and
ultimate disappearance of organs which,
after completion, fall into disuse. But it
is impossible to account, on the same
mechanical principles, for the slow but
steady building up of elaborate structures,
the functional use of which lies wholly in
the future.

The universal instincts of the human
mind are conscious that this conception is inseparable from that kind of
guidance and direction which we know
as mind. No other is conceivable.
And this particular kind of agency is as
much an object of direct perception—

when we see an elaborate apparatus growing up through many rudimentary stages to an accomplished end—as the relations of the same apparatus to the chemical and vital processes which are subordinate agencies in the result. But it is a cardinal dogma of the mechanical school that in Nature there is no mental agency except our own ; or that, if there be, it is to us as nothing, and any reference to it must be banished from what they define as science. This is all the stranger since the existence of rudimentary organs, on the way to some predestined end in various functional activities, is the universal fact governing the whole phenomena of embryology in the course of ordinary generation. Moreover, it is the very men who insist on embryology as a confirmation of their special theory, who object most vehemently to its principles being consistently applied to the explanation of kindred facts in the structure of animals in the past.

So hostile have Darwinians generally been to this interpretation of rudimentary organs in adult animals, that some years ago, when, in controversy with the late Dr. George Romanes, I spoke of rudimentary organs being interpretable sometimes "in the light of prophecy" rather than in the light of history, he challenged me to specify any one organ in any creature which must certainly have been developed long before it could have been of use. I at once cited the case of the electric organs of the Torpedo and of some other fishes. The very high specialisation of these organs, and the immense complexity of their structure, demonstrate that they must have passed through many processes of organic development before they could be used for the wonderful purpose to which, in that creature, they are actually applied. Romanes was too honest not to admit the force of the illustration when it was put before him. He took refuge in the plea that it is a solitary exception, and he declared that if there were many such structures in Nature he would "at once allow that the theory of Natural Selection would have to be discarded."[1]

Of course this plea of absolute singularity is negatived by the very first principles of biological science. There is not such a thing existing as an organ standing absolutely alone in organic nature. There are multitudes of organs very highly specialised; but there is no one which, either in respect to materials or in respect to laws of growth, is wholly separate from all others. What may seem to be

[1] *Darwin and after Darwin*, vol. i. p. 373.

singular cases are nothing but extraordinary developments of the ordinary
but exhaustless resources stored in the
original germs of all living structures.
Very special, very wonderful, and very
rare as electric organs undoubtedly are,
they do not stand alone in any one
species. They exist in fishes of widely
separated genera. Moreover, it has only
been lately discovered that they exist in
a rudimentary condition, quite divorced
as yet from functional activity, in many
species of the Rays, our own common
Skates being included in the list. Nay,
farther, it has long been known that in
all muscular action there is an electrical
discharge, so that the concentration of
the agency in a specially adapted organ,
of which we have actual examples in
every stage of preparation, is almost
certainly nothing but the development,
or the turning to special account, of an

agency which is present in all organic forms.

But this plea of Romanes, though futile as an argument for the purpose for which he used it, is at least a striking testimony to the fact that those who have been most possessed by the Darwinian hypothesis, do consider any appeal to the agency of mind as hostile to their creed.   Yet nothing can be more certain than that it is not hostile to the general idea of development, nor to the general idea of what Mr. Spencer calls organic evolution.   Provided these conceptions are so widened as to include that Agency of which all Nature is full, and without perpetual reference to which the common language of descriptive science would at once be reduced to an unintelligible jargon—provided the development, or evolution, of previsions of the future, and of provisions for it, are

fully admitted—there is no antagonism
whatever between these general conceptions and the facts of Nature.

The result of all these considerations
seems to be that when we meet with
structures in living animals, or bits of
structure, which have no function, we
never can be sure whether these represent organs which have degenerated or
organs which are waiting to be completed.
All that is certain is that they are parts
of the vertebrate Plan.   That plan has
always implicitly contained, at every
stage in the history of organic life,
elements and tendencies of growth which
must have included both true rudiments
of the future, and also real vestiges of
the past.   There is, indeed, one supposition which would put an end to our
search for organs on the way to use for
some future species—and that is the
supposition that the development of new

specific forms has, on this globe at least, been closed for ever.   I have often been amused by the smile of incredulity which comes over Darwinian faces when the very idea of the possibility of new species being yet to come, is put before them.   Yet if we had been living in the Pliocene Age—an age, comparatively speaking, very recent and of no great duration—we should undoubtedly have seen the processes in full operation by which the highest of our Mammalian forms were perfected and established. Nevertheless, the half-unconscious conviction may be true, that nothing of the same kind is going on now, and that not only has the creation of new germs been stopped, but that procreation has also been arrested in its evolutionary work.

It is curious how well this instinctive impression, which, although never expressly stated, is always silently assumed by the current assumptions of biological science, fits into the language of those "old nomadic tribes" who wrote on creation 3000 years ago, and of whose qualifications for doing so Mr. Spencer seems to speak with such complete contempt. They knew nothing of what is now technically called science. But, somehow, they had strange intuitions which have anticipated not a few of its conclusions, and some of which have a mysterious verisimilitude with suggestions which come to us from many quarters. Their idea was that with the advent of Man there has come a day of "rest" in the creative work. It does look very like it. But this supposition or assumption does not in the least affect the possible interpretation to be put upon certain rudimentary structures in existing organisms. That interpretation simply is, that the old Plan has been

followed to the last; that all the
marvellous implications and infoldings
which lay hid in the original germs have
kept on unfolding themselves—till Man
appeared.   In this case, the arrested
structures would naturally exhibit traces
of the processes which had been going
on for millions of years, although they
were now to be pursued no farther.
Thus the mere existence of a rudimentary organ, apart from other evidence,
would not of necessity imply that the
creature in which it appears is the offspring of other creatures which had that
same organ in perfection.   The alternative interpretation is easy, natural, and
may well be true—that such a rudiment
neither has ever been, nor is yet ever to
be, developed into functional activity.
It may be where it is—simply because it
indicates an original direction of growth,
or of development, which was made part

of the vertebrate Plan from the beginning of the series, for the very reason of its potential adaptability to an immense variety of purposes. Moreover, the arrest of such tendencies of growth, at a given point in the series, may well have been part of the same Plan from the beginning. But the survival of their effects—the traces of this method of operation—would thus be a perfectly intelligible fact.

As already said, the case which presents all these problems in the most striking form is the case of the Whales, and especially the case of that species which, from the commercial products of its organism, is most widely known. Both the organs which in this creature are present as rudiments alone, and those which, on the contrary, are very highly developed and most wonderfully specialised, are equally significant. Constructed

exclusively for oceanic life, it yet pos-
sesses in a rudimentary form some of the
most characteristic bones of the terrestrial
Mammalia. Upon the assumption that
no organic structure can possibly have
any other origin than ordinary generation,
and that they can never have been originated except by actual use, nor be found
incomplete except as the consequences
of disuse, then of course the conclusion
seems unavoidable that the Whale is the
lineal descendant, by ordinary generation,
of some animal that once walked upon
the land. Accordingly, I have heard a
very high authority on Biological science
declare that not only did he accept this
conclusion, but that he could conceive
no other solution of the problem presented by the facts.

Yet it is evident that it rests entirely
on the two preliminary assumptions above
specified. Of the first of these two as-

N

sumptions — that no organic structure has ever come into existence except by ordinary generation — we cannot even conceive it to be true.   But putting this aside, of the second of these two assumptions, namely, that organic structures can never have been developed except by actual use, it may be confidently said that it is certainly unfounded.   We cannot be sure that the calling into existence of new germs—a process in which the whole animal world must confessedly have begun — is a process which was adopted only once, and has never been repeated in the whole course of time. We cannot, therefore, be certain that the Cetacea, which constitute a very distinct division in the animal kingdom, have not been thus begun, with pre-determined lines and laws of growth which stand in close relation to the development of all the terrestrial Mammalia. But, even if we adopt the assumption that this alternative is impossible or inconceivable, the second assumption is certainly unjustifiable — that by the methods of ordinary generation rudimentary organs can never have arisen except by actual use, nor can have been atrophied except by subsequent disuse. The whole course of organic nature contradicts this assumption absolutely. All organs pass through rudimentary stages on their way to functional activity. And if ordinary generation has been made to do the work of forming new species, the original germs in which the process began must presumably have passed through the same characteristic steps.

The facts of Palæontology seem to indicate that the vertebrate series began with the Fish. Out of them, therefore, on the Darwinian theory of Develop-

ment, the Mammalia must have come,
and if so it is not wonderful, but quite
natural, that we should find one branch
of the Mammalian type to be organisms
pisciform in shape, and otherwise specially adapted to a marine life.   One
fundamental difference between the
Fishes and the Mammalia is in the
method and machinery for breathing, or,
in other words, for the oxygenation of
the blood.   But comparative anatomists
tell us that in Fishes the homologue of
the Mammalian lung is the membranous
sac which is called the air-bladder.   If
ordinary generation, doing nothing except what we always see it doing now,
has given birth to all creatures, it must
have done much greater marvels than
converting a mere bladder of air into a
vascular organ for mixing that air with
a circulating current of blood.   The
existence of rudiments of legs, and of a

pelvis for the support of legs, is amply
accounted for if we suppose that the
elements of the whole vertebrate Plan
were present, potentially, from the beginning of the type, with an innate tendency
to appear in embryotic indications from
time to time.   Both Owen and Mr.
Spencer, representing very different
schools of thought, have likened this
idea to that of the growth of crystals
along determinate lines, and bounded
by determinate angles.[1]   Owen goes so
far as to call the imagined initial structures by the name of "organic crystal-
lisation."   Although there is a danger
in passing, without great caution, from
the inorganic to the organic world, yet
this is a general analogy which is a real
help to thought.   The almost infinite
complication of even the simplest organic

[1] *Principles of Biology*, vol. ii. p. 8 ; Owen's
Physiology, vol. iii. p. 818.

structure when compared with the mere
aggregations characteristic of crystalline
forms, does, indeed, make it impossible
to conceive that organic growths can be,
in fundamental principle, like that of a
crystal. But in the one circumstance,
or condition of determinatedness in the
direction of growth, a common feature
may undoubtedly be recognised. It is
quite conceivable that the "physiological
units" of all organic structures should be
under the control of a force which determines their unknown movements and
mutual arrangements, so as to build up,
and form, the most complex structures
needed for future functions in distances
of time however far away. The truth
is that this conception is nothing more
than a bare description of the facts. It
supplies us with a far more simple and
conceivable explanation of the Cetacean
pelvis than the alternative suggestion

that a fully-formed land animal, with limbs completed for walking on the land, has given birth to offspring which abandoned the use of them, and acquired, by nothing but ordinary generation, all the purely marine adaptations of the Whale.

There is, perhaps, no creature so highly specialised. The baleen in the mouth is one of the most wonderful cases of an organic apparatus expressly made for one definite and very peculiar work —namely, that of forming a net or sieve for entangling and catching the millions of minute crustaceans and other organisms which swarm in the Arctic seas. It is one of the structures which classifiers call aberrant—cases in which the directive agency—so evidently supreme in all organic development—has pursued a certain line of adaptation into the rarest and most extreme conditions determined by a very peculiar food. In the pursuit of that line of adaptation it is really not much of a puzzle that one particular element in the vertebrate skeleton should be passed over and left, as it were, aside, because it is a part of the original plan which could be of no service here. There is no rational ground for supposing that this particular bit of internal structure must necessarily have been developed into functional use in some former terrestrial progenitor. Organic beings are full of structures which are variously used, and of others which are so embryonic that they can never have been of any use at all. On the other hand, it is a very violent supposition that the external structure of the Whale can ever have been inherited from a terrestrial beast by the normal processes of ordinary generation. The changes are not only too enormous in amount, but too complicated in direction, to lend

themselves to such an explanation. The
fish-like form of the whole creature—the
provision of an enormous mass of oily
fat, called blubber, completely enveloping
the internal organs, for the double purpose of protecting from cold those organs
which are dependent on a warm Mammalian blood, and of so adjusting the
specific gravity of the whole creature as
to facilitate flotation on the surface of
the ocean, where alone respiration can
be effected by the Mammalian lung—the
development of a caudal appendage which
does not represent the Mammalian tail,
but is constructed on an entirely different
type—the assigning to that tail a function
which it never serves in the Mammalia
—that of propulsion in the medium which
is its habitat—all these, together with
the baleen in the mouth, constitute an
assemblage of characters departing so
widely from the whole Mammalian class,

that if the creature possessing them has
acquired them through no other process
than ordinary descent from parents which
were terrestrial beasts, then we are
attributing to ordinary generation everything which is intelligible to us as a truly
creative power.   The stages through
which such an enormous metamorphosis
could only have been conducted, if they
were sudden and rapid, would have been
visibly a creative work ; and if they
were slow and gradual they must have
followed certain lines of growth as
steadily, as surely, and with as much
prevision, as we can conceive in any
intellectual purpose of our own.   Nothing, therefore, is gained by those who
dislike the idea of rudimentary organs
being regarded as provisions for a future
in some one original Plan, when they try
to escape from that idea by supposing
that this rudimentary condition can be

due to nothing but degeneration. That element of prevision of, and provision for, the future, which they choose to call the supernatural, pursues them through every step of their substituted fancies— and that, too, in the case of the Whales in a more immanent degree.

Mr. Spencer's tone, then, of remonstrance against the hardness of our hearts in being so slow to accept completely the teachings of the Darwinian School as an adequate explanation of the facts of Nature, shows that he has not grasped the difficulties which we feel to be insuperable. He is quite right in saying that even if the special theory of Darwin be abandoned, there would still remain to be dealt with what he calls the theory of organic evolution. Yes, and if the particular theory which he so calls be given up, there will still remain another theory which is equally entitled, and, we

think, better entitled, to the name. Let
him exhaust the meaning of his own
language. An organ is an apparatus
for the discharge of some definite vital
function. That is its only meaning. It
is a means to an end. But the existence
of a future need, and the preparation for
the supply of it, have no necessary or
merely mechanical connection. A steam-
engine must have a boiler, and a piston,
and a condenser, and gearing to convert
rectilinear into rotatory motions. These
are all needs—if the apparatus is to do
its work. But this is a great " if." For
it implies that there is some agency
which has willed and determined that
the work must and shall be done. It
implies that the mechanical needs for the
doing of it will not be supplied without
an agency which both sees them and is
able to provide for them. All vital
organs are, therefore, in the strictest

senses of the word, apparatuses, and as such are essentially purposive. The evolution of them can only mean the unfolding of elements contained in the present, but conceived and originated in the past.

We believe in organic evolution in this deepest of all senses. We do not believe, any more than Mr. Spencer, in creation without a method—in creation without a process. We accept the general idea of development as completely as Mr. Spencer does. We accept, too, the facts of organic evolution, so far as they have yet been very imperfectly discovered. Only, we insist upon it, that the whole phenomena are inexplicable except in the light of mind—that prevision of the future, and elaborate plans of structure for the fulfilment of ultimate purposes in that future, govern the whole of those phenomena from the first to the last. We insist upon it that the naked formula—now confessed to be tautological—of "survival of the fittest," is an empty phrase, explaining nothing, and only filling our mouths with the east wind.

Mr. Spencer does, indeed, towards the close of his article, use some language which may mean all that we desire to be included in the stereotyped phrase — organic evolution. He says that all the vast varieties of organic life are " parts of one vast transformation," displaying "one law and one cause," namely this, " that the Infinite and Eternal Energy has manifested itself everywhere, and always in modes ever unlike in results, but ever like in principle." But everything in this language rests on the sense in which the word Energy is here used. Etymologically, indeed, it is a splendid word, capable of the sublimest applications. We do habitually, in common

speech, apply it to the phenomena of
mind, and if we think of it in that application — as a name for the one source
from which all " work " ultimately comes
—if we think of it as that which "works"
inwardly everywhere as the cause and
source of all phenomena—then, indeed,
Mr. Spencer is making use of ideas which,
in more definite and more appropriate
language, are familiar to us all.   But,
unfortunately, the word Energy has been
of late years very largely monopolised by
the physical sciences, in which it is used
to designate an ultimate and abstract
conception of the purely physical forces.
We talk of the energy of a cannon-ball,
of the energy of an explosive mixture, of
the energy of a head of water.   We even
erect it into an abstract conception representing the total of Matter and of all
its forces, alleging that there is only a
definite sum of energy in the Universe

which can never be either increased or
diminished, but can only be redistri-
buted.   If this be the purely physical
sense in which Mr. Spencer uses the
word "energy"— even   although   he
prints it in capitals, and although he
adds the glorifying qualifications of " Infinite " and " Eternal "—then we must
part company with him altogether.   The
words "infinite" and "eternal" do not
of themselves redeem the materialism of
his conception.   The force of gravitation may be, for aught we know, infinite
in space, and eternal in duration.   But
neither this form of energy, nor any
other which belongs to the same category of the physical forces, affords the
least analogy to the kind of causation
which is conspicuous in the preconceived
Plan, in the corresponding initial structure, and in the directed development of
vital organs as apparatuses prepared

beforehand for definite functions. The force of chemical affinity is one of the most powerful of the physical energies in Nature. It is one great agent—even the main agent—in digestion. But it could neither devise nor make a stomach. Substitute for the word "energy" that other word which evidently fits better into Mr. Spencer's real thought —viz. the word "mind"—and then we can be well agreed. Then Mr. Spencer's fine sentence is but a dim and confused echo of the conception conveyed in the line so well known to most of us—"And God fulfils Himself in many ways."

Since these pages were written it has been announced that Mr. Herbert Spencer has completed the really Herculean labour of building up his "Synthetic System of Philosophy." It does not need to be one of his disciples to

join in the well-earned congratulations
which men of the most various schools
of opinion have lately addressed to a
thinker so distinguished. The attempt
to string all the beads of human knowledge on one loose-fibred thread of
thought called Evolution has been, I
think, a failure. But the beads remain,
ready for a truer arrangement, and a
better setting, in the years to come. We
must all admire the immense wealth of
learning and the immense intellectual
resources, as well as the untiring perse-
verance, which have been devoted to
this attempt. Mr. Spencer has vehemently denied that his philosophy is
materialistic. But he has denied it on
the ground that, as between Materialism
and Spiritualism, his system is neither
the one nor the other. He says expressly of his own reasonings that "their
implications are no more materialistic

than they are spiritualistic, and no more spiritualistic than they are materialistic. Any argument which is apparently furnished to either hypothesis is neutralised by as good an argument furnished to the other." This may be true of the results in his own very subtle mind, but it is certainly not true of the effect of his presentations on the minds of others. Nor is it true in the natural and only legitimate interpretation of a thousand passages.

Even in close contiguity with the above declaration of neutrality we find him asserting that "what exists in consciousness in the form of feeling is transformable into an equivalent of mechanical motion."[1] I believe this to be an entirely erroneous assertion. No calculable quantitative relation whatever has been discovered between any form

[1] *Principles of Biology*, vol. i. p. 492.

of mechanical motion and any of the phenomena of sensation or of thought. But whether this assertion be erroneous or not, it is certainly not easily to be reconciled with the claim of neutrality. An assertion that all feeling may be correlated with certain organic motions in the brain or nervous system may be true.   But that all "feeling" is "transformable into" mere mechanical motion is an assertion of the most pronounced materialism.   The truth is, that so profoundly hostile is Mr. Herbert Spencer to all readings of mental agency in natural phenomena that when his own favourite doctrine—that of evolution— gives a clear testimony in favour of such readings he not only rejects its testimony, but tries all he can to silence its very voice.

I know of no subject in which the pure idea and the pure facts of evolution open up so wide and straight an avenue into the very heart of truth as in the subject of human thought automatically evolved in the structure of human speech. Words are not made; they grow. They are unconsciously evolved. And that out of which the evolution takes place is the functional activity of the mental consciousness of Man in its contact with the phenomena of the Universe. What that consciousness sees it faithfully records in speech. It is like the highly-sensitised plates which are now exposed to the starry heavens, and which repeat, with absolute fidelity, the luminous phenomena of Space. What should we think of an astronomer who thought himself entitled to manipulate this evidence at his pleasure — to strike out appearances, however clear, which conflict with some cosmic theories of his own? Yet this is precisely the course taken by Mr. Herbert Spencer when he encounters a word which is inconsistent with his materialistic preconceptions. Although the purest processes of evolution have certainly made that word, he rules it out of court, and sets himself to devise a substitute which shall replace the mental by some purely physical image. Thus, for example, the word "adaptation" is indispensable in descriptive science. Mr. Spencer translates it, because of its implications, into the mechanical word "equilibration."[1] Thus the tearing teeth of the carnivora are to be conceived as "equilibrated" with the flesh they tear. It is curious to find Mr. Spencer thus indulging in an operation which excites all his scorn when it is resorted to by others. Adaptation is a word born of evolution. Equilibration

[1] *Principles of Biology*, vol. i. p. 466.

is a "special creation" of his own, and a very bad creation it is. Laboriously classic in its form, it is as laboriously barbarous and incompetent in its meaning. No two ideas could be more absolutely contrasted than the two which Mr. Spencer seeks to identify and confound under the cover of this hideous creation. The conception of a statical "equilibrium" or balance between opposite physical forces, and the conception of the activities of function so adjusted as to subordinate the physical forces to their own specific and often glorious work—these are conceptions wide as the poles asunder. Nothing but a systematic desire to wipe out of Nature, and out of language—which is her child and her reflected image—all her innumerable "teleological implications," can account for Mr. Spencer's continual, though futile, efforts to silence those spiritualistic readings of the world, which have been evolved in the structure of human speech.

But even if it were true that Mr. Spencer's writings are as neutral as he asserts them to be, nothing in favour of their reasonings would be gained. A philosophy which is avowedly indifferent on the most fundamental of all questions respecting the interpretation of the Universe, cannot properly be said to be a philosophy at all. Still less can it claim to be pre-eminently "synthetic." It may have made some—and even large —contributions to philosophy. But the contributions are very far indeed from having been harmonised into any consistent system. On the contrary, very often any close analysis of its language and of its highly artificial phraseology will be found to break it up into incoherent fragments. Such at least has

been my own experience; and I am glad
to think that in a line of interpretation
which leads up to no conclusion, and
to no verdict, on the one question of
deepest interest in science and philosophy—namely, whether the Physical
Forces are the masters or the servants
of that House in which we live—no
man is ever likely to succeed where
Mr. Herbert Spencer has broken
down.

THE END

Printed by R. & R. Clark, Limited, Edinburgh

The Duke of Argyll's Works.

THE BURDENS OF BELIEF

and other Poems.

Crown 8vo. 6s.

OUR RESPONSIBILITIES FOR TURKEY.

Facts and Memories of Forty Years.

Crown 8vo. 3s. 6d.

" This is a book which it does one good to read. Within the compass of 166 pages the Duke gives a panoramic view of the Eastern question from the Crimean War down to the present time, so comprehensive, yet so clear, that the unprejudiced reader is forced by the logic of facts, rather than by appeals to his emotions, to adopt the Duke's conclusions as inevitable."—*Daily Chronicle.*

IRISH NATIONALISM:

An Appeal to History.

Crown 8vo. 3s. 6d.

JOHN MURRAY, ALBEMARLE STREET.